THE POLITICS OF GENETIC RESOURCE CONTROL

Also by Tim Gray

BURKE'S DRAMATIC THEORY OF POLITICS (*with Paul Hindson*)

THE FEMINISM OF FLORA TRISTAN (*with Maître Cross*)

FREEDOM

THE POLITICAL PHILOSOPHY OF HERBERT SPENCER

THE POLITICS OF FISHING (*editor*)

UK ENVIRONMENTAL POLICY IN THE 1990s (*editor*)

The Politics of Genetic Resource Control

Anthony J. Stenson
postdoctoral student
Department of Politics
University of Newcastle

and

Tim S. Gray
Professor of Political Thought
University of Newcastle

First published in Great Britain 1999 by
MACMILLAN PRESS LTD
Houndmills, Basingstoke, Hampshire RG21 6XS and London
Companies and representatives throughout the world

A catalogue record for this book is available from the British Library.

ISBN 0–333–74502–7

 First published in the United States of America 1999 by
ST. MARTIN'S PRESS, INC.,
Scholarly and Reference Division,
175 Fifth Avenue, New York, N.Y. 10010

ISBN 0–312–22102–9

Library of Congress Cataloging-in-Publication Data
Stenson, Anthony J., 1972–
The politics of genetic resource control / Anthony J. Stenson, Tim
S. Gray.
p.   cm.
Includes bibliographical references and index.
ISBN 0–312–22102–9 (cloth)
1. Germplasm resources, Plant—Law and legislation.   2. Plant
varieties—Patents.   3. Germplasm resources, Plant—Government
policy.   I. Gray, Tim, 1942–   .   II. Title.
K3876.S74   1999
341.7'58—dc21   98–53539
                                                             CIP

This book is printed on paper suitable for recycling and made from fully managed and sustained forest sources.

10   9   8   7   6   5   4   3   2   1
08   07   06   05   04   03   02   01   00   99

Printed and bound in Great Britain by
Antony Rowe Ltd, Chippenham, Wiltshire

Contents

Acknowledgements and Dedication

This book originated in a doctoral thesis by Anthony Stenson supervised by Tim Gray at the Department of Politics, University of Newcastle. The authors wish to thank the British Academy's Humanities Research Board for the award of a research studentship to Anthony, and Newcastle University for providing resources without which this book would not have been written.

The book is dedicated to Anthony's parents, in recognition of their encouragement and support during the long years of its gestation.

Abbreviations

CBD	Convention on Biological Diversity (Rio, 1992)
CGIAR	Consultative Group for International Agricultural Research
CHM	Common Heritage of Mankind
CIMMYT	International Centre for Maize and Wheat Research
CIPRs	Community Intellectual Property Rights
CPGR	Commission on Plant Genetic Resources (FAO)
CUBE	Concertation Action in the Biotechnology Programme (EU)
EPC	European Patent Convention
EPO	European Patent Office
EU	European Union
FAO	Food and Agriculture Organization
FRs	Farmers' Rights
GATT	General Agreement on Tariffs and Trade
HYVs	High Yielding Varieties
IARCs	International Agricultural Research Centres
IBPGR	International Board for Plant Genetic Resources
IPRs	Intellectual Property Rights
IRRI	International Rice Research Institute
IUPGR	International Undertaking on Plant Genetic Resources
LGUs	Land Grant Universities
MTA	Mutual Transfer Agreement
NGO	Non-Governmental Organization
PBRs	Plant Breeders' Rights
PGRs	Plant Genetic Resources
PIPR	Plant Intellectual Property Rights
PPA	Plant Patent Act (USA, 1930)
PTO	Patent and Trademark Office
PVPA	Plant Variety Protection Act (USA, 1970)
R&D	Research and Development
RAFI	Rural Advancement Foundation International
SAESs	State Agricultural Experimental Stations

TNCs	Trans-National Corporations
TRIPs	[Agreement on] Trade-Related Intellectual Property Rights
TRRs	Traditional Resource Rights
UNCLOS	United Nations Conference on the Law of the Sea
UNCTAD	United Nations Commission on Trade and Development
UNDP	United Nations Development Programme
UNEP	United Nations Environment Programme
UNESCO	United Nations Education, Scientific and Cultural Organization
UPOV	[International] Union for the Protection of New Varieties of Plant
WIPO	World Intellectual Property Organization
WTO	World Trade Organization

1 Introducing the Politics of Genetic Resource Control

INTRODUCTION

Biotechnology has been responsible for a number of media stories in the past few years. More often than not these stories are 'novelty' stories, tales of seemingly outlandish achievements in genetic engineering such as the cloned sheep or the mice that glow in the dark (to take the two most recent examples). These stories usually give rise to a round of media moralizing, in which the question often raised is whether it is right that humanity should 'play God' with nature in this way. Biotechnology does indeed raise fundamental questions about the relationship humans have with their environment and with other living things. But there are other serious moral questions raised by biotechnology that never see the light of day in the media because of its obsession with these 'novelty' stories. These are questions of the relationships human beings have with each other.

It often seems that modern biotechnologies indicate the final step in the mastery of humans over the rest of the natural world. With genetic engineering, we can move a gene from one species to another in order to give the recipient characteristics it would never have had if natural selection had been left to its own devices. This is undoubtedly a huge advance in our power to manipulate nature. However, the idea that biotechnology represents a fundamental break from nature is mistaken. Biotechnology depends on nature completely. Genetic engineers cannot make genes. They can only take them from one organism and put them in another. The genes themselves are the product of nature, of millions of years of evolution.

Genes, then, are natural resources like any other, and as with other natural resources, there arise disputes over who ought to control – and therefore have the right to exploit – them. These disputes constitute the politics of genetic resource control. They have intensified as a result of advances in biotechnology and the concomitant realization that such resources are now potentially more valuable than ever. What makes disputes over genetic resource control particularly interesting is the fact that, on the one hand, most such resources are to be found in the developing world, yet, on the other hand, most of the scientific and technical expertise and hardware connected to biotechnology is found in the industrialized world. This has put the question of genetic resource control on to the global stage, and introduced many questions concerning the global distribution of wealth into the politics of genetic resources. The politics of genetic resource control are therefore carried on at an international level.

THE FOUR PRINCIPLES

This study is about the role of normative ideas or moral principles in the politics of genetic resource control. Its main aim is to identify, clarify, examine and appraise those ideas or principles. There are four main principles that characterize the debate about the politics of genetic resource control: proprietarian intellectual property rights; community intellectual property rights; national sovereignty; and the common heritage of mankind. Let us say a little about each of these principles in turn.

First, the principle of proprietarian IPRs. This is the prevailing principle in the debate on genetic resources today: indeed it dominates discussion. Intellectual property was included in the Uruguay Round at the insistence of the USA, who favoured its inclusion in the General Agreement on Tariffs and Trade (GATT) rather than in the more obvious forum of the World Intellectual Property Organization (WIPO) because of the cross-retaliatory measures available in GATT but not in WIPO. The questions making up the proprietarian intellectual property right principle are framed in either economistic or entitlement language. Economistic questions are concerned with which intellectual property arrangements best promote society-wide economic efficiency, while the questions framed in entitlement language interpret intellectual property in more obviously 'moral' terms, and are concerned with which arrangements are the 'right' ones.

The issue of proprietarian IPRs is where the most ferocious battles over genetic resource control have been fought. Indeed, the extension of proprietarian intellectual property rights to genetic resources can be thought of as the most important reason why genetic resource control is a salient political issue. As we describe in Chapter 2, it was only when modified germplasm, protected through intellectual property rights, began to flow from North to South that the question became a political issue. The proprietarian intellectual property right principle is primarily a western idea, adumbrated by judges, lawyers, companies and civil servants in the West. The dominant principle, or ethic, therefore, within the genetic resource control debate is the proprietarian/authorship[1] ethic, which, as we shall see, replicates the individualistic assumptions of western legal discourse.

The second principle is that of community IPRs. This principle is generally associated with the idea of the rights of indigenous communities to own the genetic resources located within their traditional habitats. In the last few decades the issue of the rights of indigenous peoples has loomed larger on the world stage. It has attracted much rhetorical support, but, inevitably, less practical support. Indigenous peoples are the custodians of many of the world's useful genetic resources, which, through selective breeding, they have often themselves improved. Indigenous peoples also hold a vast wealth of folk knowledge on the properties of wild plants and animals. As we shall see in later chapters, this knowledge is becoming increasingly sought after by Northern companies who wish to turn it into profitable products. Rights to protect this knowledge, and also traditional varieties, have been added to the list of demands made by indigenous groups in their fight for recognition and survival. The principle that has come to dominate the discourse of indigenous rights, at least with regard to genetic resources, is the principle that communities ought to be granted a form of intellectual property rights over their genetic resources. We use the term 'community intellectual property rights' to describe these arrangements, because that is simply the most accurate term to use, as will be made clear in Chapter 4. However, the reader should bear in mind that the discourses of intellectual property and of indigenous rights are distinct, and that the demand for community intellectual property rights come out of the latter rather than the former.

The third principle is that of national sovereignty. This principle lies at the foundation of international law regulating the conduct between states. It asserts that nation states have sovereign rights over the natural resources lying within their territory. Naturally these rights include genetic resources. The vast majority of the world's genetic resources are to be found in developing countries, while most of the world's biotechnical expertise and technology is to be found in the developed world. North–South politics is therefore one of the most significant influences on the politics of genetic resource control. Since the players in this politics are national governments, we identify the third principle in the politics of genetic resources as that of national sovereignty over genetic resources.

The fourth and final principle is that of the common heritage of mankind. This is the most recent principle, enunciated by Arvid Pardo, the Maltese Ambassador to the United Nations, in 1968. It asserts that humanity as a whole has a right to a share of the natural resources of the world. Despite the fragmentation of the Third World's development strategy over the past decade or so, there remains an influential strand of multilateralism, with Third World countries recognizing their common interests and acting together as a group to further these interests. In the case of genetic resources, this multilateralism manifests itself in the ethic of common heritage.[2]

CLARIFICATION OF TERMINOLOGY

We use the term 'genetic resources' rather than the more fashionable term 'biodiversity', because biodiversity is generally used in connection with discussions about effective *management* of natural resources, and tends to ignore normative questions of their *ownership*. But these normative questions must be addressed if sustainable development is to be achieved. Development is essentially human activity; all economic activity ultimately involves the exploitation of natural resources, and any use of natural resources involves assumptions over who is entitled to use resources and how this use is to be governed and by whom, i.e. questions of distributive justice.

Political ecologists might be alarmed at this portrayal of the Earth's living organisms as resources for humanity to use.

However, viewing organisms as resources does not necessarily commit one to a full-blooded human-instrumental view of nature. To talk of resources is merely to capture the sense in which biodiversity is used by people to meet their basic needs. All human life ultimately depends on using living organisms; constituting part of nature as resources for human purposes is therefore an inevitability of human existence. But this does not mean that nature must always be seen in such terms.

The term 'genetic resources' encompasses both plant and animal resources, but it is plant resources that are the most interesting from our point of view because that is where the political controversy surrounding genetic resources is centred. Animal genetic resources are not nearly so valuable as plant genetic resources, either in monetary or wider terms. So when we refer to genetic resources, it is plant genetic resources that are foremost in mind.

Another choice of terms needs to be explained before we proceed further. We refer to genetic resource *control* rather than genetic resource *ownership* because the two concepts are not synonymous: control is often exercised in ways other than through property rights. For example, national sovereignty over genetic resources is an important principle governing the control of such resources, but sovereignty is not the same as ownership. A natural resource – such as a quarry – can be owned privately while still operating under the sovereignty of the government whose territory it lies in.

OUTLINE OF CHAPTERS

Chapter 2 is a historical overview of the issue of genetic resource control, and an attempt to characterize the present situation. It describes how, over the course of the twentieth century, plant breeding has become increasingly dominated by the private sector, and claims that this process is ultimately responsible for the controversy that surrounds genetic resource control today. In terms of characterizing the present, Chapter 2 describes the reach of intellectual property with regard to plant genetic products, and also describes the effect of the principal international agreements with regard to genetic resources.

Chapter 3 discusses the dominant principle in the debate on the politics of genetic resource control – that of proprietarian intellectual property rights. We begin the chapter with a consideration of categorical objections to 'patents on life'. There is a widespread argument that it is fundamentally immoral to allow patents on living organisms. We consider it necessary to refute this belief, first because we believe it to be false, and second because denying that there are any categorical objections to life patents allows us to consistently adopt the instrumental attitude to intellectual property that we advocate in the next section of the chapter as an alternative to proprietarianism. This refutation also allows us, in the following chapter, to advocate a form of intellectual property rights for communities in their traditional varieties and botanical knowledge. The second section of Chapter 3 draws primarily on the work of Peter Drahos and James Boyle on the present intellectual property system. We describe the nature of these writers' criticisms of the system and attempt to relate these criticisms to our own case, that of genetic resources.

Chapter 4 is concerned with the question of whether indigenous and farming communities ought to be granted a form of intellectual property rights in their traditional varieties and botanical knowledge. This is one of the central claims of defenders of indigenous peoples and is one of the outstanding issues in the politics of genetic resources. This chapter is divided into three sections. The first is an introductory section, explaining the importance of traditional varieties and knowledge. The second section considers, and rejects, arguments claiming that communities are entitled to rights in their resources and knowledge. The third section develops an argument for community intellectual property rights on the basis of autonomy, and addresses some practical questions concerning such rights.

Chapter 5 deals with the principle of national sovereignty, and explains how deeply rooted this principle is in international environmental law and practice. So much so, that it is enshrined in global environmental conventions such as the Rio Convention on Biological Diversity (Article 15), where an acknowledgement is made of the 'sovereign rights' of states over their natural resources, including their rightful authority to control access to their own genetic resources (Grubb et al.,

1993, p. 79). We evaluate the ethical validity of the right of sovereign states over genetic resources in their territory; we ask why it took so long for the principle of national sovereignty to be applied to genetic resources; and we examine the relationship it has with the other three principles.

Chapter 6 considers the fourth and last principle – that of the common heritage of mankind. This principle represents the widest application of the claim to genetic resources; from the individual (proprietarian) to the group (community) to the state (national sovereignty) we come, finally, to humanity as a whole. In this chapter we begin with the notion of common heritage as expounded by Arvid Pardo in 1968, but then develop the notion, using the theory of John Rawls (1971) to deepen the ethic and give it a stronger basis. The remainder of the chapter is devoted to showing that the common heritage ethic still has an influence in the politics of genetic resource control, despite the widespread belief that it is dead, and is compatible with both the principles of community IPRs and national sovereignty.

Finally, in Chapter 7, we present a conclusion in which we summarize the main points of the thesis and its policy recommendations, and suggest some wider implications of the points we have made on the issue of genetic resource control.

NOTES

1. Drahos (1996) refers to 'proprietarianism' in intellectual property, while Boyle (1996) refers to the 'authorship myth'. As we argue in Chapter 3, both characterizations are essentially the same, despite different terminology.

2. We do not mean to suggest that the national sovereignty principle is exclusively a Northern notion, nor that the common heritage principle is only advanced and supported by Third World governments. Indeed, national sovereignty was instituted recently as a guiding principle in genetic resource control because of pressure from Third World governments, but as a general principle it underlies the whole of international relations. Common heritage is designed to raise the position of Third World countries but arises from an internationalist worldview that is supported by some western governments.

2 The Historical Context of Genetic Resource Control

INTRODUCTION

This chapter is intended to give a historical overview of the issue of genetic resource control in order to explain the current position. The chapter begins with an account of the rise of private plant breeding in the twentieth century and the concomitant entrenchment of intellectual property rights in plant genetic resources. It is the extension of intellectual property rights – in the form of plant breeders' rights and, latterly, full utility patents – in the industrialized world that fuels much of the contemporary controversy over genetic resource control.

IPRS AND THE RISE OF PRIVATE PLANT BREEDING IN THE TWENTIETH CENTURY

Over the course of the twentieth century, plant breeding has become increasingly the concern of the private, rather than the public, sector. The shift has been gradual but we are now at the stage where two-thirds of agricultural biotechnological research[1] is carried out by private capital (*Financial Times*, 30/3/95). Public sector research has been redirected away from competition with the private sector in the supply of seed to less threatening basic research. Along with this drift towards private dominance of plant breeding, the twentieth century has also witnessed a massive effort at protecting plant varieties through various forms of intellectual property rights. These trends are inextricably linked; it is our intention in this section to trace this history, and to show how the private breeding industries and, latterly, the biotechnology industries, have influenced the political and legal systems of the developed world in order to strengthen intellectual property rights in their favour. The trend towards greater intellectual property protec-

8

tion has, in the last two decades, spread to the international level; this has provoked a response from countries in the developing world, who are the original suppliers of almost all germplasm used in plant research. Traditionally, unmodified germplasm had been regarded as the 'common heritage of mankind'; the rise of intellectual property rights in modified germplasm, however, has led the developing world to reassess this situation. This reassessment has in turn led to a fundamental change in the status quo concerning genetic resources. In this chapter we will also give a historical account of the developing countries' increasing frustration on the issue from the 1970s onwards, and of their attempts to hold the growing proprietarianism of the richer countries of the world in check.

HYBRID CORN: A FOOT IN THE DOOR FOR CAPITAL

This apparent asymmetry in the ownability of germplasm (Northern researchers; Southern suppliers) has existed for some time, but has not always been a salient political issue. The controversy over ownership can, we believe, be traced back to the 1960s, with the beginnings of a flow of commercial germplasm, protected by intellectual property rights, from North to South. However, the roots of the debate go back further still, to the 1920s and the origins of the modern capitalist seed industry. Before the 1920s, seed transactions were primarily carried out by farmers themselves, although there was a minor commercial sector based around small family concerns. The state of technology and the inherent character of biological resources, however, prevented the growth of a major industry. Private firms could breed seed varieties, but once they were released, farmers could save seed from their harvest to plant in the next season. They would only have to visit the market once for every variety they wished to grow.

A technological breakthrough in the 1920s, however, opened the door to capital accumulation. This was the discovery of techniques for creating hybrid corn. From the late nineteenth century in the United States, significant amounts of public federal resources had been devoted to scientific plant breeding. Two types of organization carried out this research: so-called land grant universities (LGUs) and State Agricultural

Experimental Stations (SAESs) (Buttel and Belsky, 1987). In the 1920s these organizations successfully hybridized corn, a revolutionary breakthrough due to two characteristics of hybrids: first, hybrids are 'dwarves', displaying heterosis or hybrid vigour, which means that they grow much shorter than standard varieties, using the energy saved to produce much greater quantities of edible material. Second, this characteristic only obtains in the first generation. Plants produced from the seeds of hybrids are genetically dissimilar to the first generation and produce far inferior yields. This means that farmers must go back to their supplier every season for new seed. The advantages for the private breeder are obvious. There were other potential methods of crop improvement that might have had similar levels of success as hybrids, such as population improvement, but hybridization was chosen as the dominant method. Jack Kloppenburg believes the choice was not a matter of science but of political economy. Certainly, entrepreneurs were not slow to spot the opportunities afforded by corn hybridization. A number of companies were set up in the 1920s and 1930s to take advantage of the new technology, the most famous of which, the Hi-Bred Corn Company, is now a grain multinational (Pioneer Hi-Bred) (Kloppenburg, 1988, p. 105).

A technological breakthrough opened the door to private capital in plant breeding, but the old biological barrier to profitability – the easy reproducibility of plants – was still present in the case of most crops. However, a major theme in the history of private plant breeding has been that where technology has been inadequate in overcoming the biological barrier to profitability, capital interests have worked hard to promote the adoption of legal measures which would overcome biological obstacles to capital accumulation. As Fleising and Smart show, private interests have been able to use both lobbying power and the ideology of progress to make conditions favourable for themselves (Fleising and Smart, 1993). Even before the successful hybridization of corn, seed companies in the US had been lobbying for some time for legislation which would protect new varieties from being copied and sold without the breeder's permission. In 1930, as a result of renewed lobbying efforts inspired by the discovery of hybrid corn, Congress passed the Plant Patent Act (PPA). The PPA gave legal protection to varieties of asexually-reproducing plants and was the first intellectual property rights regime for plants anywhere in the world. Although it is doubtful whether the PPA was adequate – from the point of view of private industry – in terms of the protection it offered, or whether it stimulated innovation, it nevertheless established the principle that private plant breeders should have legal monopoly rights to the fruits of their investment (Buttel and Belsky, 1987, p. 34).

THE 'GREEN REVOLUTION'

The beginning of the North–South flow of commercial germplasm was the historical event that turned germplasm ownership into an international political issue, but this flow had been initiated earlier on by the so-called 'Green Revolution', with its emphasis on *free* exchange of seed. The green revolution was spearheaded by philanthropic organizations like the Ford and Rockefeller Foundations, and later the International Agricultural Research Centres (IARCs), which were public organizations funded by industrialized countries. The aim of the green revolution was not entirely altruistic; the self-interest of the richer countries was also at work. One aim was to use the power of capitalist technology to impress the public and raise living standards, thus averting the threat of communism and nationalism (Kloppenburg, 1988, p. 158; Oasa and Jennings, 1982, pp. 32, 39). Another aim of many of those involved in the Green Revolution was to obtain native genetic resources from the countries whose agriculture they were improving. Kloppenburg quotes Edward May, president of an Iowa seed company that was cooperating with Iowa State University, on the decision of the University to establish a corn research project in Guatemala:

> Past experience with other crops has taught us not to confine our search exclusively to our own corns. Thus it is that the Tropical Research Center has been located in Guatemala to search for genes or characters that will improve our corns and thereby contribute to greater freedom from hunger and improve the welfare and security of all nations. (In Kloppenburg, 1988, p. 159)

No payment would be made to Guatemala or Guatemalan farmers for the removal of germplasm from their lands. Such germplasm was common heritage, belonging to all mankind and available to anyone who could make good use of it. Note also how May identifies the interests of his company with the interests of mankind as a whole; free donation of germplasm is thought to be self-evidently in the interests of developing countries. At this time – the late 1940s – there was no political controversy over this situation; private breeding was not yet an international concern, since IP protection for plant varieties had not become widespread. As a result, the norm of free exchange of germplasm held sway for many more years.

Kloppenburg shows how the pattern of genetic resource transfer between North and South has been 'largely unidirectional: from the Third World to the developed nations' (1988, p. 169). This has been the case throughout modern history. In the days of colonialism the transfer took the form of what has been called the 'Columbian Exchange': crops moved from colony to colony as whole swathes of colonial land were given over to commercial crop production, but germplasm also moved from South to North as the European powers sought to use exotic genes to improve their agriculture. In the twentieth century the transfer of genetic resources has taken the form of germplasm collection and storage in 'genebanks'. Germplasm expeditions and collections began in 1898 when the US Department of Agriculture set up its 'Section of Seed and Plant Introduction'. The collections created from such expeditions were the base from which improved varieties were produced from the 1920s onwards (Kloppenburg, 1988; Fowler and Mooney, 1990). Indeed, collections of one sort or another are still the resource that scientists turn to when searching for new genes to breed or insert into their elite lines. Their value to the plant breeding industry and to humanity as a whole cannot be overstated.

The Green Revolution can be said to have begun with the Mexican Agricultural Program, inaugurated in 1943 at the instigation of the Rockefeller Foundation, but it was in the 1960s that it reached take-off point, when the Rockefeller and Ford Foundations together set up the International Rice Research Institute (IRRI) in Manila (1959), and the International Centre for Maize and Wheat Research (CIMMYT) in Mexico City (1963). A series of similar centres for different crops soon followed and collectively came to be known as the IARCs. In 1971 the Consultative Group for International Agricultural Research (CGIAR) was set up to coordinate them. At this time those involved in the Green Revolution, along with some Third World activists, became worried that the spread of the revolution, with its high-yielding but genetically uniform varieties, was eroding the genetic base of world agriculture. Genetically uniform crops carry the risk of increased vulnerability to pests and diseases; it is therefore crucial that a genetically diverse base is maintained if the collapse of whole strains is to be avoided. Only with a diverse genetic base to support it is modern monocultural agriculture sustainable. It was with these considerations in mind that the CGIAR appointed the International Board for Plant Genetic Resources (IBPGR) in 1974 to coordinate the disparate and haphazard network of germplasm collections around the world. It was imperative that these collections should not fall into a state of disrepair or confusion; collectively they held the world's food security.

The vast majority of the world's genetic diversity in food crops is to be found in the developing world. Accordingly, the majority of the germplasm found in collections originates in the developing world. Almost all of the genetic improvements made to Northern agriculture during the twentieth century have their origins in the Third World. By Kloppenburg's reckoning, exotic germplasm has been worth 'untold *billions*' of dollars to the advanced capitalist nations (1988, p. 169). Yet all such germplasm was given freely and not a penny has been earned by the developing world from it. As we have indicated, however, while the public ethos (albeit with hidden private and ideological aims) dominated plant breeding, and free exchange of seed was the norm, there was no perception of unfairness on the part of the developing world. The industrialized world might have benefited a great deal in the past from the developing world's germplasm; but these benefits were distributed publicly. Private capital had not yet – or not significantly – appropriated genetic resources for private gain.

THE RISE OF PRIVATE CAPITAL IN PLANT BREEDING

However, from the 1950s onwards this began to change. Kloppenburg describes how, by then, the private seed industry was becoming multinational in scope in the industrialized world, but found market creation a problem in the Third World – indeed, this is still a problem today. For the private seed industry to flourish in a country, there needs to be a class of farmers in whose interests it is to produce monocultural, high yielding varieties (HYVs) of crop. Such farmers must be capital-rich and technologically sophisticated in order to be able to afford and to use the fertilizers, irrigation, machinery and pesticides necessary for HYVs to achieve their full potential yield (Dahlberg, 1979, p. 45; Oasa and Jennings, 1982, p. 40). Before the Green Revolution, however, quite the opposite was the case in the developing world. Farming was carried out for subsistence rather than profit. Subsistence farming is still the majority mode of farming in the developing world even today, but it is of no use to private capital in plant breeding.

Kloppenburg believes that one of the aims of the green revolution, an aim it achieved to an impressive extent, was to create the kind of social structures in Third World agriculture that were amenable to capital accumulation:

> Though they represent a vast potential market, most farmers in the Third World have been too poor to afford commercial seed even where they are available. However, the Green Revolution has helped to galvanize the emergence of a growing class of well-capitalized and technologically sophisticated producers who are receptive to commercial seed and able to pay for them. (Kloppenburg, 1988, p. 169)

The first green revolution program, the Mexican Agricultural Program, concentrated on hybrid corn. Kloppenburg believes the emphasis on this crop is not surprising, given the 'shoe-horn' effect of hybrids in terms of the opportunities they open up for capital (Kloppenburg, 1988, p. 170). But, of course, the easy reproduction of *non*-hybrid crops means that private companies will not breed and market such crops in the absence of legal protection. The problems of easy reproducibility have been circumvented in the industrialized world, thanks to the expansion of intellectual property rights. We have already mentioned the USA Plant Patent Act of 1930; two more developments are worthy of note. In 1960 a group of European countries met to form the [International] Union for the Protection of New Varieties of Plants (known as UPOV), aimed at securing 'plant breeders' rights' (PBRs) to privately-bred varieties. These PBRs went further than the 'plant patents' granted under the American PPA of 1930; they protected all types of plant variety, including sexually-reproducing species, as long as they fulfilled the qualifications of novelty, uniformity and stability. At first, UPOV was limited to European countries, but the USA passed its own UPOV-style laws in 1970 (the Plant Variety Protection Act (PVPA)). By 1970, then, at the height of the Green Revolution, plant breeding in the industrialized world was becoming more and more dominated by private breeding, with the increased legal protection for the results of private breeding that this implies. The Third World was still 'unprotected', however; indeed, the Indian Patent Act of 1970 explicitly forbids any sort of IPR protection for agricultural and horticultural inventions (Shiva, 1996, p. 1627).

History shows that the industrialized countries chose to decide for themselves when to institute patenting. In contrast to the present, the USA at the turn of the century had extremely lax IP laws. This laxity allowed the USA to achieve competitive advantage in technology – by copying other countries' technology – much more quickly than if every piece of new imported technology had been heavily protected from copying by law. Sahai reports that Italy did not allow product patents (as opposed to process patents) until 1982, and Spain until 1987; this allowed their chemical and pharmaceutical industries to mimic foreign technology and build up an adequate technology base in order to become competitive (Sahai, 1994, p. 89). The Indian position, and that of other developing nations, on IP is therefore understandable and perhaps justifiable. But the situation in the 1970s seemed intolerable to Northern plant breeding interests – huge potential markets were denied them simply because the right legal conditions were not in place. In the 1970s, private breeders began to press for the extension of PBR legislation to the Third World (Kloppenburg, 1988, p. 170). However, the initial results of this effort were counter-productive, provoking a Third World backlash, as we shall see.

At this time, the germplasm issue was becoming politically salient, for three reasons. The first was the internationalization of the seed industry; we have already mentioned this. The second was the rapid development of biotechnology in the

1970s. Almost from the beginning, with the discovery by Stanley Cohen and Herbert Boyer of the method of splicing a DNA sequence from one organism into another, the biotech revolution was a private sector revolution. The initial advances were made in university laboratories, but almost immediately private sector capital was there to try to turn basic advances into commercial money-making practice. Indeed, three years after making the crucial breakthrough, Herbert Boyer had left the University of San Francisco and had set up Genentech, the first biotech start-up company. The biotech revolution made it clear to everyone in the developed and developing world that germplasm would soon be a very important resource; its private sector character meant that private profit would be at stake and heightened the sense of political importance surrounding it.

Shortly after the invention of rDNA, the biotech revolution had its first major legal effect. In the late seventies a molecular biologist working for General Electric in Chicago, Ananda Chakrabarty, used rDNA techniques to create a unicellular organism – a variant of the bacterium *E. coli* – capable of breaking down oil. It was thought such an invention would be useful in combatting oil slicks. Initially the patent office ruled against the patent on the grounds that the supposed invention was a living organism and therefore unpatentable; the company appealed and in 1980 the case ended up in the Supreme Court. In a revolutionary ruling, the Court declared that it was irrelevant whether an invention was living or non-living; as long as it fulfilled the usual patent criteria – novelty, utility, nonobviousness – it was patentable. The phrase actually used by the Court to describe the scope of the US patent statute was 'anything under the sun that is made by man'; full utility patent protection was extended to all inventions (Kloppenburg, 1988, pp. 261–2; Sterckx, 1997, p. 18; Hamilton, 1993, p. 596; Busch et al., 1991, p. 27). The signal that this ruling sent out was staggering in its implications. It meant there was a great deal of money to be made from the new biotechnology: living resources and their ownership were to become very important, economically (Svatos, 1997, p. 300).

The third factor leading to the rise of the germplasm issue was the growing dissatisfaction the developing world felt over its position in the world economy. The above developments in the private seed industry and in biotechnology were occurring at a time when the developing world was becoming increasingly frustrated and angry at what it perceived to be the unfair structural inequalities of the world economic system. It seemed to them that their poverty was subsidizing the developed world's wealth; they began to call for a 'New International Economic Order' which would allow them to trade on fair terms and remove some of the structural obstacles to their development. In this atmosphere the asymmetries in the world germplasm system became a serious source of tension. Encouraged by activists and authors like Pat Mooney,[2] the developing countries came to perceive the germplasm situation as grossly unfair: from their point of view, original resources had arisen, and indeed were still flowing, from the developing world to the developed world as 'common heritage' – freely available to anyone – where, thanks to superior Northern technology, they were being developed into elite strains that gave Northern agriculture a massive advantage over its Southern counterpart. It was bad enough that the developing world had never been compensated for the use of its germplasm. Now things were worse: genetic resources with their origins in the developing world were being converted into a commodity and protected by legal rights. Worse still, they had begun to be sold back to the very place they had originally come from.

The Third World decided it had had enough of this situation and in 1983, at a session of the Food and Agriculture Organization (FAO) in Rome, adopted Resolution 8/83, the International Undertaking on Plant Genetic Resources, which declared all PGRs to be 'the common heritage of mankind and should be available without restriction'. The notion of common heritage was familiar enough to everyone – PGRs had always been seen as common heritage and therefore freely exchanged. However, developing nations explicitly included in their definition of common heritage 'special genetic stocks, including elite and current breeders' lines and mutants' (FAO, 1983, p. 5). If free access was to be the prevailing norm regarding 'raw' germplasm, it was to apply to 'elite' seed too. This meant no intellectual property rights at all on any modified germplasm. As Kloppenburg says, this definition 'directly challenges the commodity form' (Kloppenburg, 1988, p. 173), but unfortunately for the developing world, it never had any chance of being adopted by the industrialized world. William Brown puts the point succinctly:

> To ask that an elite parental line which costs a company several hundred thousand dollars to develop be exchanged for cultivars of limited or unknown potential is simply not reasonable, and seed companies will not agree to such an arrangement. (Brown, 1988, p. 225)

Whether the Undertaking was morally compelling or not, there was no prospect of either the plant breeding industry or western governments ever agreeing to it. As we have shown, the trend from the 1920s onwards has been in favour of increased 'privatization' in the plant breeding sector, and concomitantly towards the greater legal protection of privately bred varieties. To suddenly declare all private varieties 'common heritage' – which did not necessarily mean common property but, to the drafters of the 1983 Undertaking, undoubtedly meant 'freely available' – was to fly in the face of the seemingly inexorable march of capital. The US, Denmark, Finland, France, West Germany, the Netherlands, Norway, Sweden, the UK and New Zealand all stated that it was not possible for them to sign the Undertaking. We believe the Undertaking is best seen as a point-making exercise on the part of Third World countries. It let the developed world know that they were not happy with the principle of common heritage governing access to their raw material, if private interests and intellectual property rights were to dominate the advanced breeding side of the equation. However, since 1983, developing countries have found that rather than directly challenge the commodity form, their best strategy has been a conciliatory one: they have accepted the trend towards privatization of PGRs and have attempted to get the best deal they can out of that situation.

PGRS PRIVATIZED: 1983 TO THE PRESENT DAY

Following the *Diamond* v. *Chakrabarty* judgment, a number of product claims on plants were made in the US. It was now legal precedent that patents could not be denied on the basis that the object of the claim was living. All claims on non-hybrid plants, however, were rejected by the Patent and Trademark

Office (PTO), on the grounds that Congress had enacted separate intellectual property legislation for plants and this legislation pre-empted patents. But in 1985 a patent was granted to Molecular Genetics on the tissue culture, whole plant and seeds of a mutant maize line selected from tissue culture to overproduce an amino acid, tryptophan (Kloppenburg, 1988, p. 263; Roberts 1996, p. 532). The PTO's Board of Appeal, following the lead of the Supreme Court in *Diamond* v. *Chakrabarty*, declared that the PPA and the PVPA did not pre-empt patents. The case, known as *ex parte Hibberd* in the legal world, completed the trend in the US toward maximum IP protection of plant innovations; plant breeders now had a choice of protection. Full utility patents for plants are now routine in the US (Roberts, 1996, p. 532). Indeed, according to Ronald Schapira (1997, p. 171), 'in the USA, nearly all products & processes of biotechnology that are *new*, that are *unobvious* and that are *useful* can be patented'.

Given the position on IPRs in the international arena today, it is astonishing to note that, just over ten years ago, patents were not available for plants in the US. The trend towards full patentability of all inventions, living and non-living, has now been completed in the international legal sphere, via the 1994 General Agreement on Tariffs and Trade (GATT). We will describe the provisions of GATT shortly, but in between *ex parte Hibberd* and GATT there were four other significant international developments in IPRs in PGRs.

Revisions to the FAO Undertaking
We earlier suggested that the 1983 Undertaking was best seen as a point-making exercise on the part of the Third World. But it did not seem like that at the time: the delegates to the FAO from the developing countries undoubtedly meant every word. There was so much rancour surrounding the Undertaking in the immediate period after it was declared, that the *Wall Street Journal* coined the term 'seed wars' to describe the situation. As we have made clear, there was emphatically never any chance of the Undertaking being adopted by the developed world. But this did not mean that the rich countries could simply ignore the Undertaking. The biotechnological revolution meant genetic resources would be of immense value in the next few decades and everyone knew this. The vast majority of the

world's biodiversity, both in crop landraces and wild germplasm, is to be found in developing countries. The developing world therefore had a potent bargaining chip – restrictions on future access to indigenous and potentially very valuable germplasm. The rich countries had an incentive to compromise.

This compromise was reached in 1989. A new interpretation of the Undertaking declared plant breeders' rights to be compatible with 'common heritage' while also recognizing the principle of 'farmers' rights', i.e., that most of the world's valuable germplasm comes from the developing world and is the result of centuries of selection by farmers, and that some form of compensation should be paid to the developing world's farming communities for these resources. The new interpretation of the Undertaking stated that farmers from the developing world had not been sufficiently rewarded, and urged the setting up of an 'International Gene Fund' to remunerate the developing world as a group for the germplasm they have given to the world community (Shand, 1991, p. 133). The fund would also be aimed at conserving the world's stock of genetic variability in its food crops. The new interpretation of the Undertaking was received favourably by the developed countries – it seems that if IPRs were not threatened they were prepared to compromise. Even the USA's position altered: it announced it would join the FAO Commission and later become a party to the Undertaking (Shand, 1991, p. 134).

Another significant alteration to the Undertaking was made in 1991. In a further move away from the principle of common heritage, the FAO adopted the principle of national sovereignty: all genetic resources were subject to the sovereignty of the country in which they were found. This meant that companies and scientists from the developed world would no longer be able to take germplasm from a country without permission or payment. As we shall see in the following section, national ownership has become an accepted principle, giving the developing world a good measure of control over their genetic resources, and it has led to a number of deals between Northern companies and developing nations.

Revisions to UPOV

If the developing world gained some concessions in the 1989 and 1991 revisions to the FAO undertaking, they gained little from the 1991 revisions to UPOV. Carlos Correa states that 'the revised Convention reflects to a large extent the wishes of large R&D companies working in modern biotechnology' (Correa, 1992, p. 155). At present, internationally, the version of UPOV negotiated in 1978 is still in force. The 1978 version of UPOV has two 'exemptions': a secondary breeders' exemption and a farmers' exemption. The first exemption enables breeders to use a protected variety as a source of variation in breeding a new variety that can itself be protected. The farmers' exemption entitles farmers to save protected seed from one harvest to plant in the next (Winter, 1992, p. 171).

However, the 1991 version of UPOV ends the farmers' exemption, at least in part. The presumption is against the farmers' privilege, and states must enact measures if they wish to allow it. Even then the right-holder may prevent such use (Correa, 1992, p. 156). This measure brings plant breeders' rights into line with patents on this issue. Under patents any replication of a protected invention without permission or royalty payment is forbidden. When plants are protected under utility patents, then, seed-saving is abolished. This is the case with the new UPOV (upon which the EU Plant Variety Right, now incorporated into UK law, is based), but there is some flexibility: states can choose to allow seed-saving if they wish. Although at present, no developing countries are actually members of UPOV, the provisions of the Agreement on Trade-Related Intellectual Property Rights (TRIPs), which we shall come to shortly, may necessitate their becoming members. Some commentators believe developing countries may be able to accede to the 1978 version of UPOV and stay out of the 1991 agreement (Verma, 1995, p. 282). However, this may prove to be over-optimistic: everything depends on whether the developed countries, particularly the United States, are happy with developing countries remaining at the 1978 levels of protection.

The Biodiversity Convention

Although the 1992 Biodiversity Convention (CBD) is aimed at the conservation or management of biodiversity, it has implications for the issue of genetic resource ownership. At the Rio Earth Summit, it was thought that economic incentives needed to be created so that developing countries would conserve their biodiversity rather than seek quick gains through deforestation and other environmentally damaging development strategies. With this end in mind the principle of national sovereignty over genetic resources was affirmed. This was a further move away from common heritage, at least in its 1983 formulation.

It is useful to distinguish between two types of fora in which the debate on genetic resource ownership has been carried out. On the one hand, there are those where the questions that pre-occupy the developing world tend to dominate the discussion. These are the UN-sponsored fora – the FAO and the 1992 UN Earth Summit at which the CBD was negotiated. On the other hand, there are the fora dominated by the interests of the developed world – the IP forum UPOV, the trade forum GATT, and the CGIAR system which is funded by the industrialized world. Over the last ten years it has been the case that in the former type of arena, the developing world has usually emerged with a deal favourable to itself, while in the latter fora, it has been less successful. Hamilton even goes so far as to say that 'we may be simultaneously nurturing two lines of conflicting international thought and law on PIPR (Plant Intellectual Property Rights)' (Hamilton, 1993, p. 613). We have seen that the revisions to the FAO Undertaking negotiated in 1989 and 1991 were to the advantage of the developing world, while UPOV 1991 reflected the interests of private breeders. As a UN-sponsored conference with the interests of the developing world very much at the forefront of debate, the Earth Summit's final outcome, the CBD, represented a good deal for the South. Not only was the national sovereignty principle affirmed, but companies utilizing developing countries' germplasm were obliged to pay royalties to the host and to transfer technology Southwards (Margulies, 1993, p. 334). The developing countries are given priority access to biotechnological products developed from their original germplasm (Article 19). Just as significantly, the Convention links intellectual property rights to the distribution of the benefits of biotechnology, stating that intellectual property rights should not run counter to the objectives of the CBD, which includes the 'fair and equitable sharing of the benefits of the utilization of genetic resources' (Article 1; Article 16 (5)). Furthermore, legislation to limit intellectual property rights in the public interest was permitted (Article 65).

TRIPs

Many of the gains made by the developing world at the Earth Summit were weakened, however, by the final GATT agreement of 1994. The so-called 'Uruguay Round' of the GATT talks lasted for seven years, partly because of the hard line taken by the United States on intellectual property rights. The final agreement on Trade-Related Intellectual Property Rights (TRIPs) reflected this hard line. It requires countries to allow patents on all inventions (Article 27 (1)). Countries may exclude plants and animals from patentability if they wish, but they must institute some form of protection for plants, if not patents then a *sui generis* system like PBRs (Article 27 (3); this may, as we indicated earlier, mean accession to either the 1978 or 1991 versions of UPOV). Furthermore, this clause is to be reviewed after four years of the World Trade Organization (WTO) coming into existence; after this time, countries may lose the right to exclude plants from full patentability (Verma, 1995, p. 281; da Costa E. Silva, 1995, p. 547; Shiva, 1996, p. 1626).

Hence, merely ten years after patenting first became available in the most proprietarian system in the world (the USA), it was to be enforced everywhere, with few concessions. It was the USA's recognition of national interest, coupled with its threat power, that was behind the move towards maximum intellectual property protection in GATT. Intellectual property had not previously been considered a subject of GATT, but the industrialized countries, particularly the United States, had pressed for its inclusion in the talks because of the cross-retaliatory measures available through GATT – which were not available in the more obvious fora for dealing with intellectual property, the World Intellectual Property Organization (WIPO) or the United Nations Commission on Trade and Development (UNCTAD) (Khor Kok Peng, 1990, p. 210). For example, in 1996, the United States government was engaged in dispute settlement procedures with India. The outcome of the Uruguay Round gave India until 2005 to protect pharmaceutical patents; however, the USA wanted India to fulfil this obligation immediately. The Indians refused, and the USA threatened trade sanctions under the so-called 'Super 301' provisions of the 1988 Omnibus Trade and Competitiveness Act if it did not get its way (Jayaraman, 1996, p. 182).

As D.M. Nachane writes, 'the final outcome [of GATT] represents a definitive victory for the industrialized countries' viewpoint' (Nachane, 1995, p. 257). We cannot help but feel that this outcome was more or less inevitable. The developing countries' viewpoint on IPRs and genetic resources has usually won in those fora connected to the UN, but has invariably lost when the debate has been shifted to international trade organizations. The UN is a body committed to ideals of global fairness and equity but has little real power. Trade organizations like GATT/WTO and those dedicated to coordinating intellectual property rights, like UPOV, have the weight of powerful national and business interests behind them and are always more likely to achieve their aims. As Martin Khor Kok Peng wrote in 1990, well before the conclusion of the Uruguay Round, 'The Uruguay Round can be seen as the TNC Empire's grand way of striking back at global demands for legislation to tackle ecological concerns and at the Third World's demands for global economic justice' (Khor Kok Peng, 1990, p. 213).

LEGAL MOVES TOWARDS GREATER PROTECTION

It is the theme of this chapter that, as plant breeding has become increasingly dominated by the private sector in the twentieth century, the intellectual property protection afforded to plant varieties and the products of biotechnologies has increased with it. The more obvious moves towards greater intellectual property protection have been statutory: from the 1930 PPA to TRIPs, governments in the developed world have, one way or another, enacted laws and made multilateral agreements to increase protectability. Western legal systems have also played a part in this drift towards proprietarianism by interpreting existing law in favour of private interests. This process is made clear by the recent history of US and European legal judgments on biotechnology patent applications. We have already mentioned the 1980 *Diamond* v. *Chakrabarty* decision by the US Supreme Court. This was the first time a patent had been granted anywhere on a living organism, but there had been no change in the law to facilitate this decision; the Supreme Court simply chose to interpret the existing laws in a way that favoured the applicant. Although the Supreme Court is no doubt aloof from such calculation, the decision had the effect of stimulating US investment in a potentially lucrative field in which intellectual property protection is crucial in gaining adequate returns. If the decision had gone the other way, and patentability had been denied to the products of rDNA technology, investment would perhaps have gone elsewhere, and potentially billions of dollars would have been lost to the USA. Similarly, the *ex parte Hibberd* decision of 1985 was not based on any change in statute, but on a conscious decision on the part of the Board of Appeals to reverse the traditional practice that patents should not be granted on plants because there was separate intellectual property legislation covering that sector.

It is not only the US that has performed legal turnarounds to increase the protection afforded to the products of biotechnology. Most European countries, including the UK, are signatories to the European Patent Convention (EPC), which, on the basis of a single application, hands out a 'European patent', which is in fact a set of individual country patents administered and enforced by the signatory. The European Patent Office (EPO), based in Munich, awards a patent which is valid for all countries designated by the applicant, unless the courts in the individual countries disagree with the EPO Appeals Board. The EPO is presently granting patents for mainly microbiological products and processes of biotech, to the extent that Hans-Rainer Jaenichen and Andreas Schrell can declare that 'the patenting of biotechnological inventions has become routine' (Jaenichen and Schrell, 1993b, p. 466). Some patents have also been granted for multicellular, 'higher' organisms, such as the 'Harvard Onco-mouse'[3] which was granted patent in 1990 (Jaenichen and Schrell, 1993a, p. 345). Although this is, of course, not a plant variety, the Onco-mouse case is particularly interesting because it shows how the courts can and do interpret the law on patentability of living organisms in favour of private industry. Section 53 (b) of the European Patent Convention states that a Patent shall not be granted

> for any variety of animal or plant or any essentially microbiological process for the production of animals or plants, not being a microbiological process or the product of such a process. (Reid, 1993, p. 301)

One might reasonably assume that such a prohibition meant that patents could not be granted for animals or plants, and that therefore the Onco-mouse could not be patented. However, the crucial word is 'variety'. Patents shall not be granted for any *variety* of plant or animal, but this does not mean that animals and plants *per se* are to be excluded.

The initial lay reaction to such a decision is perhaps one of incredulity. How could a court claim that the Onco-mouse is not an animal variety? And if it is not an animal variety, what exactly is the patent claim for? The answer is that animals developed by non-biological and microbiological processes are not regarded as varieties and are technically patentable. Of course, throughout most of the history of the patent system, animals could only be developed through biological techniques; so the very purpose of explicitly excluding biologically developed organisms was to *ex*clude multicellular organisms but to *in*clude microbiological inventions, which have been patentable since the 1920s. The word 'biological' in patent law is historically defined in opposition to 'microbiological'; it has been taken to mean 'conventionally bred', i.e., with animals, through ordinary selective mating. The advent of biotechnology, however, means that a plausible argument can be put forward that animals can be developed using non-biological, i.e. (bio)technological, techniques. Here, 'biological' is defined in opposition to 'technological'. The crucial stage in the development of the Onco-mouse was the (technological) insertion of genes into the embryo. It was declared by the EPO to be a technological invention, and therefore not a variety, and therefore patentable.[4]

The situation is similar with regard to plant breeding. In one sense, things here are clearer. Whereas there is no accepted definition of an animal variety, the evolution of the PBR system over the previous few decades has given legal systems a definition to work with. The 1991 UPOV defines a plant variety as 'a plant grouping within a single botanical taxon of the lowest known rank' (Phillips and Firth, 1994, p. 355). Even though this definition is less rigid than previous UPOV definitions, its purpose is still clear: to exclude from protection plants developed by breeders that are not sufficiently distinct as varieties. However, the definition of a plant variety given by UPOV is now being used to *in*clude plants in the patent system

– which is a stronger form of protection than the PBR system. In 1990 the Oppositions Board[5] of the EPO granted the company Lubrizol a patent on what to the lay person seems like a plant variety, on the basis that it *lacked* stability in some trait of the whole population (stability is a requirement for varieties to be protected under PBRs) (Reid, 1993, p. 22). The motive of the Oppositions Board was clear and stated – they did not feel that plant variety protection was available or sufficient for botanical inventions in the biotechnological era, and that patent protection was therefore desirable (Jaenichen and Schrell, 1993b, p. 467). This is a clear statement that the legal systems of the developed world have deliberately interpreted the law in a way antithetical to the intentions of the original drafters of the law, in order to allow biotechnological inventions 'in' to the patent system; or, as Fleising and Smart put it, allowing the products of biotechnology to become commodities (1993, p. 43).

The EPO has now adopted the practice of allowing patent claims on groups larger than plant varieties (Jaenichen and Schrell, 1993b, p. 467), despite the fact that the definition of a plant variety contained in UPOV and all plant variety protection legislation was intended to exclude such large categories from legal protection. In the case of Lubrizol, it was held that the claim referred to just such a larger category and was therefore entitled to patent protection (Jaenichen and Schrell, 1993b, p. 468; Reid, 1993, pp. 22–3). Quite evidently, the EPO is doing all it can to bring biotechnological products under patent protection.[6]

The rise of private plant breeding, and more recently, the rise of biotechnology, have gone hand in hand with a radical extension and strengthening of intellectual property rights in the twentieth century. This process is most marked in the developed world, where a thoroughgoing proprietarianism has clearly come to dominate thinking about intellectual property in both the political and legal spheres. The principle that has traditionally weakened proprietarianism in intellectual property – that of balancing the interests of the potential IP holder with those of society at large – has been cast aside in favour of an ethic of 'to the property holder, all the benefits shall go'. Furthermore, the globalization of plant breeding together with the biotechnological revolution have shifted this process into the international arena, and to the developing world. Nevertheless, the developing countries have achieved some relative successes, most notably the affirmation of national sovereignty over genetic resources. Moreover, the common heritage ideal still exerts some influence on international arrangements concerning genetic resources. Although the trend towards maximum intellectual property right protection is very powerful, these other principles, as we will explain below, remain significant and could ameliorate the worst effects of global proprietarianism.

CONCLUSION

Globally speaking, the question of germplasm ownership is far from settled. Although the intensity of the controversy in governmental circles has subsided considerably since the 1980s, there still exist disagreements on many issues, such as 'Farmers' Rights', the rights of indigenous peoples in genetic resources, and the sharing of the benefits of the utilization of genetic resources. In addition to this ongoing debate at government level, a number of activists, commentators and ordinary people in both the North and the South continue to criticize and agitate against the status quo.

Without a doubt, the most significant current trend in international thinking on germplasm ownership is the move towards strong intellectual property protection for modified germplasm. However, this does not mean that Northern transnational corporations have won a total victory. National ownership over genetic resources, at least that which resides *in situ*, is by now an entrenched principle. Furthermore, as we will argue in Chapter 6, the common heritage ideal continues to exert an influence, with Farmers' Rights and benefit-sharing still a contentious and unsettled issue. There is also a heavy weight of opinion and activism behind calls for the recognition of the rights of indigenous peoples in their botanical resources and knowledge.

Having described the history of the issue and attempted to give some idea of the present situation in the politics of genetic resource control, we now move on to discuss the four principles we identified in the introduction as making up the politics of genetic resource control. First, in the next chapter, we look at the principle of proprietarian intellectual property rights.

NOTES

1. Of course, not all agricultural biotechnology is plant breeding, but the agricultural biotech sector is an historical continuation of the plant breeding industry. 'Traditional' plant breeding is still carried out, but biotechnological techniques become more preponderant as time goes by.

2. Mooney published *Seeds of the Earth* in 1979; it was an attack on the growing proprietarianism of plant breeders and a warning that the Earth's genetic resources were being destroyed by short-sighted development aimed at profit.

3. The Onco-mouse is a mouse used in medical research, bred for its high susceptibility to cancer.

4. W.R. Cornish believes this to be a justified decision. He claims that the European Patent Convention is clearly not intended to prohibit patents on animals as such, because it refers to 'animal varieties' and then to 'biological processes for the production of plants or animals'. The crucial fact is that the article refers differentially to 'animals' and 'animal varieties'. He writes: 'In using the different terms 'animal varieties' ('*races animales*', '*Tierarten*') and 'animals' ('*animaux*', '*Tiere*') in this way, the legislators cannot have meant animals in both cases' (Cornish, 1996, p. 75). However, it is unlikely that the drafters of the Convention would have seen any difference between animals and animal varieties at the time; it is only with the advent of biotechnology that it has become clear that there could be a difference, relevant to patent law, between animal varieties and animals *per se*. Cornish elsewhere shows that new intellectual property rights have been extended to new areas throughout their history, usually through a process of accretion – redefining existing rights to encompass new material (Cornish, 1993, p. 54). This process has been followed in the case of biotechnology patents. There is nothing wrong *per se* with this accretive process; however, at least in the case of biotech patents, the process has been carried out in order to improve the environment for private business to make profits out of biotechnology.

5. The Oppositions Board is the committee of judges charged with adjudicating on formal objections to patents.

6. This does not mean that the EPO is prepared to let any application pass. In 1995, for example, the Appeals Board of the EPO upheld a Greenpeace objection to a Plant Genetic Sciences/Biogen patent on the basis that 6 of the 44 claims involved in the patent application referred to plant varieties. They also argued that just because an invention involved a microbiological process, this did not make it a microbiological invention and therefore patentable (Llewelyn, 1995, p. 510).

3 Proprietarian Intellectual Property Rights

INTRODUCTION

In Chapter 1 we argued that there are four main principles now running through debates and policy on genetic resource control. The most significant of these is the proprietarian intellectual property principle. The last decade has seen an unprecedented extension of intellectual property both in terms of its geographical reach and in terms of the extension of patents into new types of organism. In this chapter we analyse and evaluate the driving principle behind these developments – that of proprietarian intellectual property rights.

The chapter is divided into two sections. In Section I we address the issue of 'patenting life'. The practice of granting patents on living organisms has attracted considerable attention; many individuals and groups have attacked the practice as immoral. We argue that this is mistaken, and that there are no categorical objections to patenting living things that do not count as objections to owning actual living things. We then consider consequential objections to 'life patenting'. In Section II we explain the thinking behind the present intellectual property system. This section is principally derived from the work of two writers, Peter Drahos and James Boyle. Finally, in Section II, following Drahos, we advocate the switch from a proprietarian to an instrumental approach to intellectual property.

SECTION I: OBJECTIONS TO PATENTING LIFE

Categorical Objections to Patenting Life

Intellectual property rights have been available on plant varieties in one form or another for some considerable time in the West without causing much controversy. As we saw in Chapter 2, the first country to inaugurate plant variety rights was the United States; the 1930 Plant Patent Act granted a limited monopoly on sexually reproducing crops. In Europe, the UPOV system, to which most European countries adhere, has been in existence for nearly forty years. However, relatively recent legal and political decisions allowing full patents to be granted on living matter have generated heated debate and much opposition. The media have carried the debate enthusiastically and campaigning groups have been set up to lobby against 'patenting on life'. Why are the public so concerned about patents when they seemed to ignore plant variety protection, which is, after all, simply another form of intellectual property ownership, for decades? There are, we believe, three reasons. The first is the rise of the new biotechnologies, and in particular of genetic engineering. Genetic engineering worries many people because it is seen as interfering in nature to an unprecedented, unwise and possibly unethical degree. It is seen as a qualitative advance in refashioning nature according to human wishes. Worries about 'owning life' and 'making life' have become bound up with each other; both are concerns about the extent of human control over nature.

The second reason is that plant variety rights are, obviously, only available for plants; most people probably do not have any strong feelings about the ethical status of plants and are not particularly concerned that certain types of plant can come under proprietary control through intellectual property rights. However, patents are available for all kinds of living matter, from genes to actual types of animals. Patents extend the area of ownership in both directions, down to the basic 'building blocks' of organisms and simultaneously up to sentient creatures. People who are not worried about the ownership of types of plants have become concerned about the ownership of molecules and of animals.

The third reason, we suggest, is that patents are more extensive in the control they give the proprietor than are plant variety rights. Plant variety rights offer the proprietor a monopoly over the sale of a certain type of plant or seed, but they allow farmers to save seed from one harvest to the next, and they allow other breeders to create new varieties that differ only slightly from the original, without having to pay royalties. The rights they offer are in fact quite limited. By contrast, patents give an unrestricted monopoly on the commercial exploitation of something for 20 years (following the GATT agreement of 1994). Patents on plants permit neither 'seed-saving'[1] nor unauthorized use of the protected variety in the development of new strains. The extent of the patent right is such that opponents can talk rhetorically of 'owning life' more convincingly than would have been possible with plant variety rights.

Many people object to patents on organic material – whether whole animals or plants, or genes, or gene fragments – on categorical grounds. Such objectors do not disapprove of patents on life simply because they think such patents will have bad consequences; they believe there to be a first-order moral objection to the practice. We believe such objections to be based on conceptual misunderstandings to do with the meaning of property rights and the practice of patenting.

The Genetics Forum is a UK organization set up to monitor developments in genetic engineering and to campaign against the extension of patenting to its products. In its publication *The Case Against Patents in Genetic Engineering*, the Genetics Forum offers an objection to patents on living things which can be seen as the basis of all categorical objections to life patenting. It argues that:

> The very notion that someone is able to own a plant or animal intellectually, to own its whole being and purpose, is morally repugnant. (Genetics Forum, 1996, p. 17)

The Forum also argues that patenting living matter entails 'treating life itself as a mere commodity' (Genetics Forum, 1996, p. 3). This complaint is echoed by Shiva who believe that patenting leads to the devaluation of life by 'reducing' it to its constituents and allowing it to be owned as private property. Patents on life are 'the ultimate privatisation of life itself' (Shiva, 1991, p. 2746).

There are a number of assumptions made by the Genetics Forum and by most of those who argue that patenting life is morally wrong, which we believe are false. In our view, there are no good objections to the practice of protecting with intellectual property rights innovations that relate to living matter that are not consequentialist in nature. The assumptions made by the anti-life-patenting lobby are threefold: first, that ownership entails domination; second, that intellectual property rights somehow entail the ownership of the very essence of a thing; and third, that patenting living things is a significant change in the relationship between humans and the rest of nature. We believe that these assumptions are false and are based on misunderstandings relating to the nature of property, and particularly intellectual property. The best way to make clear these misunderstandings is to explain what property rights actually are.

Following Wesley Hohfeld (1919), we can say that property is a relationship between persons concerning a thing. It allows one person certain rights concerning the thing in question, and denies those rights to all others. If I own something, what that means is that I am entitled to use that thing in certain ways and others are not so entitled. This analysis is very pertinent to our discussion because it shows that the important fact about property is not what it says about the relationship between owner and owned, but the way it denotes the rights and non-rights of owner and non-owner. Indeed, the property right in itself does not say *anything* about the object of the property right.

Of course, there are moral restrictions on the range of objects that are permitted to be the object of property rights. The obvious example of something that is not admissible as property is a person. This was not always the case, but nowadays most people believe that there is something about human beings that acts as a moral bar on them becoming the property of someone else – even if they have consented to it. What is being claimed, by opponents of patenting on living organisms, is that *life* itself is a characteristic that certain things have that should prohibit their ownership. If we own life, we are engaging in the same kind of morally repugnant domination that slave owners are engaging in.

The obvious problem with this argument is that people have owned living things since time immemorial. The first living things to be owned were, perhaps, hunting dogs; later, as agriculture spread, crop plants would have been the subject of property rights. As Stephen Crespi points out, 'living things can indeed be commodities, such as yeast for bakers and brewers and edible plants and animals for farmers' (Crespi, 1995, p. 431). Unless opponents of life want to condemn this universal human practice, they must change focus (Roberts, 1994, p. 371). Some people do indeed condemn the practice of human owner-ship of nature, but these people – generally deep greens or radical environmentalists – are on the fringes of ethical and political discourse. We doubt if most of the opponents of life patenting would want to follow them.

The argument used by opponents of patenting living matter to counter this point is that intellectual property is different from ordinary property: while individual organisms have been owned for a long time, patents confer ownership of a species or variety of organism and therefore represent a qualitative dis-tinction. Similarly, when the Genetics Forum talks of patents involving the ownership of a plant or animal's 'whole being and purpose', they appear to be implying that when someone owns a pet dog, they do not own that dog's 'whole being and purpose', but presumably something else – something less than its whole being and purpose, so that this kind of ownership is not morally significant – connected to it.

It is true that there is a qualitative distinction between intel-lectual property and ordinary, tangible property: whereas ordi-nary property rights confer ownership of a concrete thing, inasmuch as patents confer ownership of a thing at all, that thing is abstract, an idea. However, it is not clear that this distinction makes a difference. Intuitively, it does not seem to improve the opponent's case: why should we be concerned about someone owning an idea if we are not concerned about their owning a real, living being with feelings and a life of its own? An intellectual property right may extend the range of influence of the property right, giving the holder a measure of (a different type of) control over a more extensive area than an ordinary property right (Drahos, 1996, p. 161) (although this statement must be qualified – see below), but this seems more of a quantitative than a qualitative difference; in which case the argument against patenting life becomes consequentialist rather than categorical.

In any case, the assertion that patents confer ownership of a species or variety is questionable. If the argument is that a patent on a plant confers ownership of every single instance of that variety, this is plainly false. Patents do nothing of the sort. The actual embodiments of that type of plant – the plants – belong to whoever paid for the seed and grew them. But what of the metaphysical assumptions behind the objections to life-patenting that we referred to earlier? One of these assumptions was that patenting conferred the ownership of an organism's essence, or its 'whole being and purpose'. To demonstrate the falsity of this idea, we must look again at the meaning of a property right.

The seminal work of Antony Honoré (1961) demonstrated that what we call a property right is in fact a collection or bundle of different rights, some of which may be present or absent in each particular case. According to Honoré, there are eleven 'property rights' in all: the right to possess, to use, to manage, to receive an income, to consume or destroy, to modify, to alienate, to transmit, and to security, along with an absence of term and liability to execution. So what a property right crucially is, is a set of rights that enable you to *do* things with an object. In the case of patents, that object, if we choose to see it that way, is abstract. This means that certain of the rights we associate with property are not applicable, such as the right to possess and to destroy. There is also no absence of term with intellectual property rights: patents are limited to 20 years (following the GATT/WTO agreement of 1994). A patent consists of two essential rights: to commercially exploit an abstract object, and to exclude others from doing the same. If patents are analysed in this way, the argument that patents denote the ownership of the essence of an organism, 'its whole being and purpose', seems bizarre. Patents merely give exclusive rights to commercially exploit a thing. How does this entail the ownership of something's whole being?

The other main categorical argument against the patenting of living organisms is that patenting entails the domination of the patented organism. The moral force behind this complaint is that ownership entails domination and that it is unacceptable for humans to dominate 'life', to parcel it out amongst themselves. But we have seen that a patent denotes the 'ownership' only of an abstract object. Two points follow from this fact. The first is that 'life' is never patented: abstract objects are not themselves alive, even if they relate to objects that are. This blunts the force of the 'no patents on life' slogan. The second point is that it seems doubtful that something that only exists as a category in people's minds can be the object of human domination.

However, the claim made by opponents of patents on organic matter might be that it is not the abstract object but 'nature' that is the object of domination: when patents spread to 'life', nature is colonized. Angus Wells, for example, writes that when patents are extended to living things, 'life is simply another morsel of techno-culture to be created and consumed' (Wells, 1994, p. 116). This is similar to, and is probably influenced by, the theories of post-structural commentators like Paul Rabinow and Arturo Escobar. They argue that biotechnology and the property rights that go with it represent the incursion of culture into nature, or the 'enculturation of nature', as natural processes become controlled by humanity at the most fundamental levels (Rabinow, 1992, p. 241; Escobar, 1996, pp. 60–1).

However, political ecologists are making a mistake if they believe that nature is somehow dominated by such a process. Post-structural commentators deny the existence of an objective nature, out there and independent of human ideas about it. They see nature as one more human discourse. What they mean when they say that nature is 'enculturated' is simply that a change in human perceptions of nature is taking place, that the discourses of culture and nature are merging as a result of changes in human technology. There is no sense in which nature is being wronged by such a process, or even that it could be.

Most political ecologists, however, do believe in an independently existing nature. Bill McKibben, for example, defines nature as whatever has not been modified at the hands of humans (this is partly why he proclaims the 'end of nature', as the results of human activity, such as global warming, ensure that no part of the biosphere is left unchanged by human hand) (McKibben, 1989, p. 48). Furthermore, ecologists believe that nature has an intrinsic value, quite apart from how humans value it. They oppose patenting on living matter because it is seen as treating nature simply as a resource, ignoring the innate value it has in and of itself.

However, even if nature is an independently-existing entity, patents on parts of it – parts that have been adapted by humans and that would not occur in nature – do not imply its domination at the hands of humanity. Once again, we return to what patenting actually is. It is the right to produce something commercially while simultaneously excluding others from doing the same, for a temporary period. The only difference between the production of patented food crops and non-patented food crops is a relationship between people: in the former, certain people are excluded from the practice; the practice toward nature is identical. The production of crops is a universal human practice; it does not entail treating nature 'merely' as a resource, or as a 'mere commodity'. The only significant relationship that arises from patenting is that between people. That is because it is a legal instrument delineating the economic activities of individuals.

This analysis clearly indicates that any objections to the patenting of living matter should be (a) anthropocentric and probably (b) consequentialist but not categorical.[2] Patenting does not signify a change in the way humans treat nature, but only a change in the way they organize themselves. Patents do not signify the increased domination of nature at the hands of humanity, or ownership of the essence of a living thing.

Consequential Objections to Patenting Life

However, just because there are no good categorical objections to the patenting of life this does not necessarily mean there are no problems whatsoever with the practice. And it certainly does *not* follow that, as Crespi claims, 'patenting, as such, is not an act of moral significance' (Crespi, 1996, p. 382). As Boyle points out, '[the] rules of intellectual property may push research in a particular direction, simply because the likely result would be more easily protected than the result of some other equally promising line of research' (Boyle, 1996, p. 13). In other words, patenting has consequences, some of which may be undesirable. This alone makes it an act of moral significance.

There are many consequentialist arguments directed against the patenting of living things. Four such objections are found in the Genetics Forum pamphlet we mentioned earlier. The first is that animal welfare may be put at risk by the practice of patenting animals. So far, only one animal has been the subject of a successful patent application. This was the Harvard Onco-mouse, whose case was discussed in Chapter 2. In fact, the patent refers to any non-human mammal containing the gene that gives the Onco-mouse its unique cancer-susceptible qualities. Whatever one's views on the rightness or wrongness of animal experimentation, the Onco-mouse was clearly bred in order to suffer. To grant a patent on its invention is to encourage the breeding of more such hapless creatures, and thereby to increase, perhaps temporarily, the amount of suffering in the world. On discussing the acceptability of the Onco-mouse in terms of the morality clause of the European Patent Convention, the European Patent Office decided, in utilitarian fashion, that the potential benefits to humanity of the Onco-mouse outweighed the suffering of the mice themselves. By weighing the mice's suffering, it signalled that patent law can take account of moral considerations, and that these moral considerations include human dealings with the rest of nature. Nevertheless, it granted the patent and must therefore take responsibility for the adverse effects on animal welfare that will result.

Second, there is good reason to believe that allowing the patenting of living organisms will contribute to the spread of genetic uniformity in agriculture and the concomitant loss of crop biodiversity. This process has been going on since the beginnings of European colonialism, but it has accelerated rapidly since 1945 with the Green Revolution, which was essentially the application of modern plant breeding, based on Mendelian genetics, on a global scale. The high-yielding varieties of modern plant breeding are genetically uniform. Too high a level of genetic uniformity is dangerous because it carries with it increased susceptibility to diseases and pests. This vulnerability has been exposed in devastating fashion at various times in recent history, and not just in the developing world. The US corn crop, for example, was nearly destroyed in the early 1970s by Southern Corn Leaf Blight. Farmers across the US were growing the same type of corn which had been bred on a very narrow genetic base; when the disease came along, they were all equally susceptible. The disease was eventually eradicated by the introduction of an anti-Leaf Blight gene taken from an Ethiopian traditional variety.

In the last decade or so there has been an extensive international effort to conserve the world's crop diversity. There now exists a global network of genebanks devoted to collecting and conserving traditional varieties. Most of this conservation takes place *ex situ*, in deep freezes. But everyone recognizes that *in situ* conservation is equally, if not more, important. It is essential that traditional varieties continue to be grown so that the farmers' knowledge that is connected to them lives on. A preserved seed without the knowledge of its properties might not be useless, but clearly it is better if this knowledge lives on. However, there are good reasons for believing that the patenting of elite varieties contributes to the dying away of traditional farming. First, patents, just like plant variety rights, reward genetic uniformity by their very being. Inventions have to be described, and to be adequately described, plants must be uniform. Second, the difficulties and costs associated with the enforcement of patents will make it likely that the companies who hold the patents on genetically engineered plants will be reluctant to deal with small farmers, preferring to deal with large 'agri-businesses' whose activities can be easily monitored. Traditional varieties are generally grown by small farmers, but if they are unable to compete with agribusiness because they are denied access to high-yielding varieties, they may be forced out of business or bought out by the larger companies who are only interested in growing elite varieties. Third, and this is a point related to the previous one, for all their qualities, traditional varieties do not generally yield enough to compete with the elite varieties that are protected by patents. Farmers make money from yield; therefore, without special measures, they may be forced to abandon traditional varieties and grow elite varieties instead.

The third consequentialist argument against patents on living organisms is related to the general trend towards private sector funding of public research in Northern universities. As we saw in Chapter 2, lobbying from private sector capital led to the end of the US public breeding programs based on free distribution. Public institutions are now generally used by private sector capital as centres of basic research. Rather than pay for basic research out of their own pockets, private companies involved in plant breeding now tend to share the costs with the taxpayer through public institutions. The money provided by private capital proves useful to the continually cash-strapped universities; they are happy to take it. The deal is generally that any commercially viable products that arise out of such partnerships are patentable by the company concerned. Academic research has traditionally proceeded according to the ideals of free and unrestricted dissemination of information. However, a requirement of patenting is that the invention must be new and non-obvious. Courts in the developed world take any prior publication or public dissemination of information pertaining to the invention as evidence that the invention is not new.[3] The danger is, then, that university-capital partnerships might lead to information being suppressed in order not to jeopardize patentability. This represents a clear break with the ideals of public research and academic progress that have dominated academia for centuries. Those who care about public life in liberal democracies will be worried about such developments.

The fourth and final consequential argument against the patenting of living organisms is that it carries the danger that it will worsen the position of the developing world in relation to the North. Due to the concentration of research into the genetic engineering of patentable crop plants in the North, what is happening is that germplasm that originated in the South is being transformed by Northern businesses into patentable matter and sold back to the South. Unlike plant variety rights, patents prohibit the replanting of seeds saved from the previous year's harvest. Farmers will therefore have to buy seeds every year or pay royalties on saved seed. This represents yet another addition to the debt burden of the South.

We have seen, then, that although there are no satisfactory categorical objections to patenting living organisms, there are many potentially adverse consequences of this practice. However, as we shall see in the next section of this chapter, present patenting practice is dominated by an attitude of 'proprietarianism' that ignores the wider social consequences of patenting, concentrating only on the benefits to the patent-holder. The solution to this problem is, as Drahos argues, to abandon proprietarianism and adopt an instrumental attitude towards intellectual property that can take account of consequences.

SECTION II: THE SHORTCOMINGS OF INTELLECTUAL PROPERTY

As we argued at the end of the previous chapter, the move towards maximum intellectual property protection for modified genetic resources has come about in large part as a result of the dominance in industrialized countries, particularly the United States, of a particular way of thinking about intellectual property. Peter Drahos identifies this way of thinking as 'proprietarianism'. However, as James Boyle argues, the problem is not simply one of the present interpretation of intellectual property. There are certain biases in the very basis of Western conceptions of intellectual property that blind it, first, to some forms of information production, and second, to some of the negative effects of intellectual property. This chapter concentrates on the shortcomings of the dominant approach to intellectual property in the West, with special attention to the consequences of this approach for the politics of genetic resource control. It draws heavily on the work of Drahos and Boyle, as these two authors offer the most extensive and convincing critiques of intellectual property. Their critiques are couched in different terms and on certain points seem to be at odds. However, we believe that they are ultimately saying the same thing about intellectual property and its shortcomings. We begin this section by giving an account of Boyle's critique of intellectual property.

Boyle: Intellectual Property and the Authorship Myth

Boyle's central worry about intellectual property is its tendency to ignore certain forms of information production. In particular, 'sources' of information, which can range from the human body to indigenous community knowledge, and from which intellectual property is created, are not taken into consideration by intellectual property law. The danger of this is that the 'public domain' of freely available information may wither away. This is regrettable from everyone's point of view, particularly those who use the public domain as a source of information from which they can create new advances in knowledge.

The intellectual property system's general blindness to sources is, Boyle believes, the result of the romantic myth of the author, and the entitlement theory of which this figure forms the central focus. The myth of the author is that 'the author takes facts, genre, and language from the public domain, works on them, adds the originality of spirit presumptively conferred on him by the themes of romanticism, and produces a finished work. The ideas ... return to the public domain, thus enriching it for future use. But because the author's originality has marked the form of the work as "unique", the form or expression becomes his alone' (Boyle, 1996, p. 98).

This 'authorship myth' is not confined to works of literature and the intellectual property form that goes with them – copyright. It has an influence across intellectual property law in all its forms. A good example which Boyle gives of the influence of the authorship myth in patent law is the famous case of John Moore's cell-line. Virus-infected cells taken from the spleen of a California man suffering from leukaemia were the subject of a patent awarded to the scientists who developed the resulting cell-line. Moore disputed the patent – arguing that if anyone should hold the patent, it ought to be him – but the California supreme court upheld it. They argued that the cell-line could not have been Moore's property because his cells were merely raw materials, while the scientists had displayed 'inventive effort' in turning these cells into a cell-line (Boyle, 1996, p. 106). Boyle sums up his anxiety about the influence of the authorship myth on patent law as follows:

> Viewed through the lens of authorship, Moore's claim appears to be a dangerous attempt to privatize the public domain and to inhibit research. The scientists, however, with their transformative, Faustian artistry, fit the model of original, creative labour. For them, property rights are necessary to *encourage* research. Concern with the public domain fades away as if it had never existed. (Boyle, 1996, p. 107)

Boyle regards the authorship myth as globally unfair in that it favours the type of information production dominant in the developed world at the expense of those more common in the developing world: 'Curare, batik, myths, and the dance "lambada" flow out of developing countries, unprotected by intellectual property rights, while Prozac, Levis, Grisham, and the movie *Lambada!* flow in – protected by a suite of intellectual property laws, which in turn are backed by the threat of trade sanctions' (Boyle, 1996, p. 125). The latter creations are favoured because they can be sensibly described as the product of a deliberate creative effort by one person or a group of persons. The former cultural phenomena are types of information but are not protected because they cannot be traced to any one 'author' figure.

It might be argued that the authorship myth is useful because it rewards innovation while preventing information that is not the result of any creative endeavour – such as John Moore's genetic code – from being protected by intellectual property rights. However, Boyle argues that the system based on the authorship myth is unjustified even – perhaps especially – from this utilitarian point of view. Excessive attention to the authorship myth is damaging because it diminishes the importance of the public domain. As we indicated before, the preservation of the public domain is crucial; ignoring it can have adverse effects on further innovation, and potentially disastrous real-life effects for humanity.

To show how this is so, Boyle uses the example of the rosy periwinkle. This is a plant, used in the treatment of diabetes by the indigenous people of Madagascar, that was 'discovered' by the Lilly pharmaceutical company and which is now used to produce an anti-cancer drug with a worldwide trade worth $100 million per year (Boyle, 1996, p. 128). The drug is, obviously, the subject of a patent owned by Lilly, but the blindness of the intellectual property system to sources – in this case, the indigenous knowledge that led to research on the plant and the genetic information encoded in the plant's DNA – mean that Madagascar has never received a penny. It might be argued that there is no reason to give Madagascar a penny, given that they did not engage in any creative effort. However, the potential effects of the blindness of intellectual property systems to informational sources suggest that this is a self-defeating attitude. Madagascar is a poor country without many sources of foreign exchange. A form of intellectual property right on the periwinkle could have brought in relatively huge amounts of revenue. As it is, neither the indigenous peoples nor the government get any revenue, and deforestation and slash-and-burn agriculture continue. This carries with it the danger of cultural knowledge dying out, along with the wealth of genetic information contained in rainforests. This is clearly a massive loss to the public domain and potential future innovation. Boyle writes: 'I think the system undermines the very goal it claims to promote. The distributional inequities are purchased at the cost of *inefficiency*, rather than efficiency. The rosy periwinkle story is one that precisely exemplifies the utilitarian failures of the current regime' (Boyle, 1996, p. 140).

Drahos: the Dominance of Proprietarianism

If Boyle sees theories of entitlement and the authorship myth as the root of intellectual property's failures, Drahos identifies as the villain of the piece a particular attitude within modern thinking about intellectual property. He calls this attitude 'proprietarianism'.

Proprietarianism is a simple concept with, in Drahos' words, a 'core of distinctive normative beliefs' (Drahos, 1996, p. 203). According to Drahos, it has three basic features.

Property Fundamentalism
This feature of proprietarianism refers to the belief that property rights are normatively prior to any other rights and are therefore inviolable. This attitude might be underpinned by a natural rights ethic: entitlement theorists such as Robert Nozick (1974), for example, believe that people have rights and all that matters, normatively speaking, is that these rights are respected and protected. However, a natural rights ethic is not essential to the proprietarian attitude towards property. Indeed, as Drahos points out, Jeremy Bentham, who believed all talk of natural rights to be 'nonsense upon stilts', himself thought property rights ought to be considered inviolable. This was perhaps understandable: Bentham believed progress to be dependent on the security that private property brought about. If property owners were continually having their property rights interfered with, they would not invest in the future and progress would be jeopardized.

Clearly, then, it is possible to have a consequentialist and yet inflexible attitude to property rights. Whether this position is philosophically consistent is open to question, but this is not Drahos' point; his point is that proprietarianism is an attitude that is having real-world effects now. As long as consequentialists hold to property fundamentalism, they are proprietarians, whether or not they are consistent in so doing.

The First Connection Thesis
The general formulation of the first connection thesis is this: the person first connected to an object that has economic value, or an activity that produces economic value, is entitled to a property right in it. The use of the word 'entitled', should not

be taken to imply a natural rights foundationalism; a consequentialist could argue for the first connection principle on the
grounds that it leads to the best outcomes. In Drahos' words,
the first connection is a 'personal act of demarcation' (Drahos,
1996, p. 201). The Lockean metaphor would describe it as an
act of labour mixing, while in Hegelian terms it is personality
imprinting itself on the object of proprietorial desire. Basically,
says Drahos, the first connection is an act of control. He gives a
number of examples:

> The trapping of animals, the spearing of whales, finding a
> plant variety, the mining of the sea-bed, discovering land or
> resources in it, placing a satellite in orbit, synthesizing deriva-
> tives of penicillin, locating a gene and using the electro-
> magnetic spectrum are all examples of acts of control that
> may give rise to property rights under first possession rules.
> (Drahos, 1996, p. 202)

Negative Community
This feature of proprietarianism is much more complex than
the previous two. For this reason our discussion of it will be
more extensive. Drahos believes that any theory of intellectual
property must make use of some model of an 'intellectual
commons'. Essentially, Drahos' 'intellectual commons' is the
same as Boyle's 'public domain'. Both refer to the set of usable
abstract objects in existence:

> The intellectual commons is a resource, a resource which
> consists of abstract objects. Every community has to make de-
> cisions about the use of this resource. There is no escaping
> this. (Drahos, 1996, p. 59)

The idea of a commons is familiar to anyone who has studied
theories of property in general; it refers to any resource which
is available for use by a particular group of people. There are
two important decisions concerning how the commons is to be
conceived. First, the group that is entitled to use the commons
can be constituted in one of two ways. The group can be inclus-
ive or exclusive. This means that either the whole of humanity
can be entitled to use a particular resource – say, the fisheries
of the high seas – or access can be limited to a certain group. In
the latter case, the right to use a commons might depend on

residence in a particular village or membership of a particular cultural community.

Second, in constituting the commons, members also face the question of how appropriation is to be allowed to take place. As with the question of access, there are essentially two choices here. The commons can exist either under conditions of 'positive community' or of 'negative community'.[4] Under positive community, the commons is thought to belong to everyone; it is therefore unprivatizable. The standard example is the English commons. In every village in England, there used to be (and often still is) a tract of common land that was available for every farmer to use to graze livestock on. However, no farmer could fence off part of the commons and declare it to be their own exclusive patch. Under conditions of negative community, however, this would be possible, as the commons is available for private appropriation. Under negative community, the commons belongs to no one; everyone (as long as they are entitled to membership by whatever criteria hold) is entitled to take pieces and declare them private property. Sticking to the example of land, a perfect example of negative community would be the 'land-grabs' that were staged in the USA in the middle part of the last century. Huge swathes of land were divided up and offered for private appropriation to the first person to arrive in each division and drive their stake into the ground.

The idea of the commons is comparatively easy to grasp when it comes to physical objects like land. But what does it mean when applied to abstract objects? Drahos is keen to point out that the idea of abstract objects does not necessarily imply adherence to some kind of Platonic realm of forms. 'The intellectual property system may reject the ontological reality of abstract objects but retain the category as a convenient fiction to be used in making decisions about relations between actors' (Drahos, 1996, p. 153). Indeed, says Drahos, all theories of intellectual property, in fact all cultures, need some conception of the intellectual commons. On a 'strong justificatory theory' of intellectual property, in which labour simply creates a permanent pre-legal property right in the created object, intellectual property would be granted simply through the first connection thesis. On this view, abstract objects do not exist before appropriation; a creator creates the abstract object and is therefore entitled to property rights in it under first connection principles. But in developing this theory, the proprietarian could go one of two ways: he could say, first, that all intellectual property rights are permanent – in which case, as Drahos points out, there would never be any sort of intellectual commons because all abstract objects would be owned. However, this hard line proprietarianism is rarely advocated; most proprietarians accept the second position – that since creators rely on a society conducive to creativity, under entitlement principles the abstract object in question must revert back into public ownership. In this case, there would be an intellectual commons, but one made up only of previously owned abstract objects, and existing under conditions of positive community.

Boyle and Drahos: are the Two Critiques Compatible?

On a certain reading of these respective writers' analyses, there appear to be significant disparities between them. We can see this if we examine the contrast between what Boyle identifies as the need for an author figure at the centre of most intellectual property forms, and the 'first connection thesis' that Drahos identifies as one of the three component parts of proprietarianism. These ideas are clearly similar in many ways. The first connection thesis of intellectual property grants intellectual property rights in an abstract object on the basis of, to use Drahos' words, 'some personal act of demarcation'. Clearly, it is necessary to have an identifiable subject – or group of subjects – capable of performing this act. This subject can be thought of as Boyle's author figure. On the first connection thesis, for example, intellectual property rights can be granted to an individual who develops a new type of wheat, but not to a tribe who possess a traditional variety of wheat. Drahos would say that this is because there is no personal act of demarcation in the latter case; Boyle would say that it is because there is no identifiable author figure or figures. Drahos concentrates on the act, while Boyle concentrates on the actor, but both are saying essentially the same thing.

Drahos' idea of proprietarianism, however, does not seem to rely so obviously on the 'author-vision'. One of the points Drahos wants to make is that the proprietarian attitude seeks to grant intellectual property rights not only in 'original' abstract objects but in discoveries as well. He develops the idea of the first connection thesis partly because, he says, it legitimates intellectual property rights in discoveries as well as 'creations'. It is important for Drahos to make this point because he wants to explain certain trends in intellectual property, particularly in the areas of biotechnology, towards granting patents more and more to discoveries. Boyle's analysis, on the other hand, seems to be quite different. He argues that discoveries are one of those kinds of information production that are generally *ignored* by the intellectual property system, because of the author-vision. Because of its bias towards the romantic vision of creativity and originality, the system tends towards the protection of 'original' or 'created' information and not 'discovered' information (Boyle, 1996, p. 156). Boyle and Drahos, then, seem to be in disagreement as to the empirical reality of the intellectual property system; the authorship myth and the first connection thesis are invoked by the respective authors to explain characteristics of the system that, apparently, cannot both exist. Either the intellectual property system is rewarding discovery or ignoring it; it is either driven by the first connection thesis or by the authorship myth.

In fact, this disagreement is more apparent than real. Boyle's main grievance is not that discoveries made by western corporate science are not being protected – they clearly are – but that the system rewards only certain types of discovery, i.e., those that fit the authorship myth. This might seem counterintuitive: if the authorship myth is one that privileges 'originality' over discovery, how can it account for intellectual property rights in *any* type of discovery? But this is to look at the situation too simplistically. The authorship myth does not just legitimize intellectual property rights in original works, but, as we saw above, has an influence that extends all the way across the intellectual property spectrum. Its effect is to reward the kind of information production that results from endeavours in which an 'author figure' (or, in Drahos' terms, a 'personal act of demarcation') can be identified, and to discriminate against information that does not result from such endeavours. In the case of western corporate scientists discovering something (a substance, say, or a gene), the 'author-shaped gap' can clearly be filled. It is a deliberate effort by a team of scientists possessing 'Faustian artistry' to bring certain information into the light. The information provided by indigenous groups about the pharmaceutical properties of plants, or the information contained in the genetic code of a traditional crop variety, does not fit this bill and is therefore not protected. It does not fit the bill because in Boyle's terminology there is no 'author figure'; in Drahos' terminology, there is no act of demarcation and it does not qualify under the first connection principle either.

Essentially, then, the authorship myth and the first connection thesis are the same analysis of intellectual property systems, couched in different terms. We prefer Boyle's analysis for its ease of understanding and the fact that it does not rely on a complex theory of the intellectual commons/public domain. However, Drahos describes the negative *effects* of intellectual property in a much deeper and theoretically satisfying way than does Boyle. It is to these effects that we now turn.

Intellectual Property's Liberty-Damaging Qualities

Proprietarianism cannot be understood unless the threats that are posed by intellectual property are themselves understood. Drahos' objection to proprietarianism in intellectual property is that it fosters an uncritical acceptance of intellectual property, presenting it as inviolable, and not taking external social costs into account. It therefore encourages complacency towards the negative effects of intellectual property. In Drahos' view, intellectual property has a number of potentially negative consequences for society that can only be avoided if a more flexible, instrumental attitude to property is adopted. We find Drahos' worries convincing, but at times he seems to exaggerate the possible detrimental effects of intellectual property rights.

Essentially, the potential harmful effects Drahos identifies in intellectual property boil down to the claim that the characteristics of intellectual property are such that they can easily lead to social inequality and a loss of liberty for individual citizens. This claim is most extensively worked out in Chapter 7 of Drahos' book, on the power of abstract objects. It is therefore worth looking at the claims of this chapter in detail.

Drahos begins the chapter by seeking to establish certain facts about property *per se*, not just intellectual property; in particular, he wants to show that property can be seen as a form of sovereignty over others. Property owners, he says, direct the conduct of non-owners in a manner analogous to that of sovereigns. Making use of Hohfeld's analysis of the elements of property rights, Drahos says that '[t]he rights, powers, privileges and immunities of the property holder can be thought of as being in microcosm the structural analogue of the sovereign's power to regulate conduct through commands, prohibitions and permissions' (Drahos, 1996, p. 148). Moreover, private property is not just private power, but also public power. It is private in the sense that it is a 'single-place relation', existing between two assignable individuals, but it is also public in the sense of existing against any number of persons. A property right in land, for example, can be considered a 'right to exclude the world'.

Only under certain conditions is this sovereign-like characteristic of property significant. In some cases it does not seem to result in any important imbalances of power between individuals. If one person owns a lawnmower and another does not, for example, we would not conclude that the lawnmower-owner is more powerful in society than the other. However, writes Drahos, under certain conditions property acts as a mechanism that concentrates power to produce imbalances in power between individuals in society. In short, property can have a 'sovereignty effect' (Drahos, 1996, p. 150). It is Drahos' assertion that intellectual property acts to produce a sovereign effect and therefore to produce power imbalances in society. How does this occur? In Drahos' words, 'how is it that property in abstract objects promotes the concentration of power amongst individuals within a society rather than its diffusion?' (Drahos, 1996, p. 158).

The bulk of the answer to this question, according to Drahos, lies in the fact that abstract objects are essentially indeterminate. They have to be 'created', in the sense that their boundaries cannot be a matter of fact, at any rate to the same degree as those of physical property, but must depend on the judgments of a legal elite. This argument is related to Drahos' wider claim that the extent of the distribution of capital in society is one factor to take into account when considering the distribution of power in that society. Generally speaking, the more capital a person or organization controls, the more powerful they are. In the late twentieth century, capital takes many other forms besides concrete goods. Behind many concrete goods are abstract objects which have themselves now become capital goods, as they are the subject of ownership and control through intellectual property law. This means that the law, when it creates abstract objects, through the creation of new forms of intellectual property, or the extension of the range of protectable subject matter (through legal judgments), 'in effect creates capital' (Drahos, 1996, p. 158).

In Drahos' opinion, however, capital in abstract objects is dangerous. This is because it relates to a potentially infinite store of concrete capital. It is indeed a 'gateway' to physical capital. When there is a high level of social reliance on the physical objects to which property rights in abstract objects relate, a great deal of power is channelled to the property-holder, who acquires a sovereignty effect. Drahos explains that there are some objects upon which there is a significant degree of dependence, and relationships of dependence create conditions conducive to the making of credible threats by one party against the other; they create conditions allowing A to make use of 'threat power' against B. In order to harness this power, A has to employ some mechanism; property is an example of such a mechanism. Intellectual property – property in abstract objects – potentially leads to the accretion of vast threat power in the hands of the few. It is this fact that is the basis for Drahos' claims regarding the potentially negative effects of intellectual property on citizens' liberty. Drahos' assertions here need further explanation, since they are crucial to his argument.

When property in abstract objects relates to physical resources upon which there is a high level of collective reliance, there are two consequences. The first, claims Drahos, is that new kinds of threat power emerge that are extensive and potentially global. All property rights set up patterns of interference in society in that they prescribe limits on the behaviour of others. If I own a car, for example, that means no one else is permitted to use it without my permission. This interference is not particularly serious, however, given that most people in the world would not be in a position to use my car anyway. It would be stretching credulity to say that I was interfering in their lives when in reality my ownership of my car does not make any difference to them. However, *intellectual* property rights set up much more extensive patterns of interference in people's behaviour. If I have an intellectual property right in a particular type of car, no one can produce – and therefore, indirectly, use – any example of that car without my permission. I am interfering in many more people's behaviour. This is what Drahos calls the capability-inhibiting quality of intellectual property rights.

The second consequence of having intellectual property rights in resources on which there is a high degree of collective reliance, according to Drahos, is that power will come to be increasingly unequally distributed as intellectual property rights also are. Following Marx, Drahos argues that each successive generation of technology is, in real terms, more expensive than the last. The ownership of the relevant abstract objects 'in certain exotic areas of science' (Drahos, 1996, p. 161) is therefore only available to an ever-decreasing handful of major players. This is a vicious circle: the best abstract objects are those that lead to the costly advances in technology that allow concrete goods to be produced more efficiently. The advantages of abstract objects therefore accrue in the hands of the already capital-rich who can afford to implement them. The profits are then used in research and development to improve technology still further (by producing ideas for, for example, better machinery; in other words, by producing abstract objects).

This unequal distribution of abstract objects – and, as a result, of material resources and wealth – has far-reaching consequences for the distribution of power in society. Drahos argues that 'where those abstract objects are gateways to universally important resources it follows that proprietors of those objects acquire vast threat power' (Drahos, 1996, p. 161). For example, the owners of a block of land and a pharmaceutical company with a patent on a particular drug both have a sovereignty mechanism at their disposal, but only the patent of the pharmaceutical company has a sovereignty effect. There are invariably many blocks of land available to buyers; but there may not be an alternative to the drug owned by the pharmaceutical company. It therefore has threat power where the landowner does not. Drahos issues a warning based on such claims:

> extensive, possibly global, power will probably be concentrated in the hands of those who, through their

scientific/technological capabilities and superior capital resources, are able to capture, through the property mechanism for abstract objects, resources upon which there is a
universal reliance. (Drahos, 1996, p. 161)

Drahos has in mind such resources as genes, seeds, plants and
animal-biotechnological products which have only recently been
accorded strong intellectual property protection. His warnings
have an echo in Michele Svatos' doubts concerning the ability of
patents to stimulate competition. Patents may in certain circumstances inhibit competition; for example, the more patents
there are in a given area, the more difficult it is to avoid
infringement. This leaves the way open for large corporations
with the financial clout to be able to push for broad claims to
gain control of strategic patents and thereby dominate a particular sector of industry (Svatos, 1996, p. 130).

There are three ways in which this concentration of threat
power can have deleterious effects on liberty. First, it can
threaten political liberty. 'Republicans', argues Drahos, ought
to be concerned at the power imbalances that can be caused by
intellectual property, because 'the concentration of global
threat power in the hands of an elite...would make it harder
for a society to achieve the republican ideal of liberty, an ideal
in which citizens are part of a free society and have their interests safeguarded by the rule of law' (Drahos, 1996, p. 161; see
also Boyle, 1996, p. 32).

Second, threat power linked to intellectual property can also
damage economic liberty. Drahos refers to Friedrich von Hayek,
who defended laissez-faire economies by saying they were committed to the presence of competition and therefore prevented
the accretion of economic power into the hands of any one individual or institution. In saying this, Hayek (1944) was criticizing collectivist economies, but Drahos claims that 'private
sovereigns' are just as dangerous from this point of view:

States that enact property forms that enable private
sovereigns to harness enormous threat power embark on a
dangerous strategy, for they increase the capacity of those
private sovereigns to discipline markets and to plan against
competition. (Drahos, 1996, p. 163)

We earlier paid attention to Drahos' theory that intellectual
property rights set up patterns of interference in society that

are more extensive than those that ordinary property rights set up. This reflects the third way in which intellectual property limits liberty. Intellectual property – particularly copyright and trademark – has a capability-inhibiting quality, limiting access to knowledge and other kinds of capital that are crucial to the development of individuals and their capabilities. This is an objection to the potential power of intellectual property rights to impair a person's autonomy. Writers like Joseph Raz (1986) and Will Kymlicka (1989) value liberty insofar as it enables people to achieve their full potential, to achieve autonomy as individuals.[5] The learning acquired by individuals through being exposed to information is clearly crucial to this process. Intellectual property rights act as a bar to information and could therefore be seen as acting as a bar to people achieving autonomy. If liberty has value insofar as it allows individuals to become autonomous, we will regard these effects of intellectual property rights very seriously. To prevent people from developing their autonomous capacities amounts to the same thing as limiting their liberty.

How compelling are these arguments of Drahos? We earlier suggested that Drahos' warnings on the dangers of intellectual property are a little overstated. We will now explain why. We accept Drahos' dire warnings with regard to certain situations. When a pharmaceutical company has a patent on a drug for which there is no analogue, it clearly has an absolute monopoly and all the threat power that goes along with that. It can demand virtually any price it likes because the threat of withdrawal is so powerful. However, most intellectual property rights relate to things for which there are potential replacements. Even drugs often – perhaps usually – have potential replacements which do the same job.[6] The presence of even one competitor in a market considerably reduces the threat power of either one of the producers. In the case of the example of the car, although I may own the rights to a particular kind of car, this is unlikely to give me huge amounts of threat power because rival models will always exist. Hence, although Drahos is right to warn us that in certain situations we must be careful, given the broad patents granted in the field of biotechnology, in most cases, we believe, Drahos' warning is exaggerated.

Intellectual Property and Global Injustice

In the eighth chapter of his book, Drahos changes focus to consider the relations between property and information from the perspective of justice. The theory of justice he adopts is the Rawlsian contractarian theory; he shows that under the Rawlsian theory, certain arrangements for intellectual property, both domestically and internationally, would be ruled out. At this point Drahos introduces what he calls a 'terminological shift'. He begins to talk of 'information', rather than abstract objects. This, he says, is because of the expansiveness of the term: 'Information is the daily lifeblood of human agents as communicating beings. Talking about information rather than abstract objects captures the pervasive effect that intellectual property rights in information can have on the daily lives of people' (Drahos, 1996, p. 171). This switch allows Drahos to situate intellectual property more comfortably in Rawlsian discourse. Information, he argues, is a primary good. In Rawls' theory, primary goods are those goods that are needed by any human being if they are to pursue their own projects in life. Rawls' own list of these primary goods was short and sparse, and did not include information (1971, p. 62). According to Drahos, however, information ought to be on the list because it is necessary for planning. Planning plays an important role in Rawls' theory: humans are conceived of as beings who live their lives according to a plan.

Rawls' theory of justice is a contractarian theory: it is worked out on the basis of intersubjective agreement between individuals taking part in a hypothetical, pre-social, 'original position'. The parties in the original position are denied certain types of information about themselves, such as where they will stand on the income scale of the society to which they are to belong, or what ethnic group they will belong to. When parties are denied the information they could have used in order to slant social arrangements in their favour, Rawls believes, just arrangements and institutions will follow. However, some knowledge is obviously possessed by the parties to the original contract. For example, the contractors are capable of rational planning – they must be in order to make a contract at all – and they must therefore know that they are rational planners. To Drahos this is significant:

> Given that individuals know that they will be the rational for-
> mulators of plans, it is also likely that they will want some
> basic level of information and access to information as one of
> their primary goods. (Drahos, 1996, p. 174)

Having argued that information can be seen as a primary good, Drahos goes on to discuss how Rawls' theory of justice might be applied to it. However, this is not easy: information is an awkward concept that does not fit simply into Rawls' scheme. On the one hand, it can be thought of as a resource and there-fore subject to Rawls' 'difference principle' (part of the second principle of justice),[7] to be distributed in such a way as to maxi-mize the position of the least well-off in society. Drahos argues that the implication here is that we should exercise caution in extending the term of intellectual property rights: 'Unless there are some very clear-cut gains to the least advantaged the difference principle cuts across increasing temporal bars to information' (Drahos, 1996, p. 177).

However, information can also be thought of as a basic liberty which therefore comes under Rawls' first principle of justice. In that case, it seems, it would not be permitted to distribute it unequally, even under difference principle rules, but would have to be equally distributed as one of the basic liberties of society. Any intellectual property rights would therefore seem to involve a breach of equal liberty in that they would grant privileged access to some people to certain types of information; the lexical priority of the first principle over the second would disallow this.[8]

The fact that information can be interpreted as belonging to either the first or the second of Rawls' principles of justice high-lights, according to Drahos, the artificiality of the simple division between political and civil society that the two princi-ples of justice imply. 'The important point,' he writes, 'is that information is a primary good that has a place in both political and civil society, that is in both of Rawls' principles' (Drahos, 1996, p. 177). So how *do* we conceptualize information and order its distribution in society? Drahos offers no explicit answer, but seems to favour viewing information as a good to be distributed under the difference principle. This gives him the flexible, instrumental approach to property he needs to combat the inflexibility of proprietarianism.

We believe that there is a more satisfactory way of solving the problem. Drahos argues that the question of information shows how artificial is Rawls' division between civil society and the political sphere. However, we believe this same division could help us in working out how information actually fits in to Rawls' scheme. Freedom of information is indeed a basic liberty, but it does not apply to all types of information. When we think of freedom of information, we ordinarily think of types of information held by political authorities and used to make decisions that affect our lives. To use the phrase 'freedom of information' to advocate the freedom to obtain commercial information held by private firms – unless it is personal information – would seem strange. This is because freedom of information is a basic political liberty, which applies to the political sphere but not to civil society. Intellectual property rights apply to commerce and therefore do not come under the same rules of access that we apply to political information. The type of information owned by commercial institutions is not (overtly) political in nature; it would therefore be inappropriate to argue that it ought to be covered by the basic liberty of freedom of information in the same way that information held by political authorities is. To summarize, we can say this: commercial information relates to civil society and therefore comes under the distributive constraints of the difference principle; information held by political authorities relates to the political sphere and should not therefore be unequally distributed.

We have, then, explained how information could fit into Rawls' theory of justice but not how the ownership of information would be conceived under the original position. To explain this, Drahos introduces the concept of human capital. Human capital is knowledge and skills embodied in people. It is foundational for economic development. The question of how human capital and intellectual property link up, says Drahos, 'is rather a big question and nothing like an answer can be given here' (1996, p. 179). But there is a clear overlap between them, given that human capital is embodied knowledge and intellectual property is rights to that knowledge. Drahos suggests a number of possible adverse outcomes for society in terms of human capital as a result of intellectual property. One is that because intellectual property allows a price to be charged for knowledge, individuals' decisions over whether to add to their stock of knowledge will be affected by that price. In this way intellectual property might adversely affect the society-wide growth of human capital and also limit people's opportunities. Another possible outcome is that the more that people are encouraged to attach a price to their knowledge and skills, the less these knowledge and skills will diffuse throughout society: 'A society with a highly developed intellectual property consciousness may find it is encouraging the underexploitation of its knowledge and human capital' (Drahos, 1996, p. 180). Lastly, according to Drahos, intellectual property, through copyright, adversely affects the spread of published knowledge throughout society. At present most western societies recognize this fact and accordingly give the education sector preferential treatment *vis-à-vis* copyright.

In terms of the Rawlsian framework, these possible adverse consequences suggested by Drahos 'give rational actors in the original position another reason for treading cautiously in the design of intellectual property rights' (Drahos, 1996, p. 180). There is even more need for caution, according to Drahos, when we turn our attention to the international sphere and the issue of 'global information justice'. Drahos' main point here is that internationalizing a strong set of intellectual property compromises the ability of sovereign states to adjust their own set of property rights as they see fit. For this reason such an arrangement would be rejected by the parties to the international original position, because the power to adjust property rights is one of the most important powers a sovereign state has. The starting point in explaining why this is the case is Rawls' affirmation of an instrumental attitude towards all forms of property. We will discuss this in more detail later; suffice it to say here that an instrumental attitude is one which does not see property rights as inviolable or inflexible, but as a set of arrangements that may be used in achieving justice. For Drahos' purposes here, that means that arrangements for property rights, including intellectual property rights, would be one of the things discussed and decided upon by contractors in the original position.

Whether Rawls' theory of justice is applicable on an international level is a matter of some controversy in political theory; however, Drahos does not enter these debates because, he believes, he does not need to. He merely makes use of Rawls' own 'Law of Peoples' in order to show that a strong international intellectual property system is inconsistent with the theory of justice. The 'Law of Peoples' is the phrase Rawls uses to denote the principles regulating interaction between states. They are chosen by representatives of the world's states in a hypothetical international original position. The same kind of constraints apply in this situation as apply under the domestic original position: representatives are denied all information relating to their state that might lead them to slant arrangements in their favour. In such a situation, writes Drahos, 'certain international arrangements for intellectual property would be inconsistent with those principles that were chosen to underwrite international law' (Drahos, 1996, p. 186). This is primarily because of the potential power that intellectual property arrangements have to affect states and their citizens, and the subsequent concern the participants in the international original position would have regarding them. It is also because the world is heterogeneous, and even under Rawlsian procedures there would likely be a variety of local conceptions of justice. Participants would therefore have to devise arrangements whereby these local variants could survive in the world.

For this reason, according to Drahos, non-interference would be one of the norms chosen, and any arrangement concerning intellectual property that threatened extensive interference in the capacity of local cultures to support their variants of justice-as-fairness would be rejected. This means that a 'globalized protectionist scheme' of intellectual property would be ruled out. A 'globalized protectionist' scheme would have four features: it would favour longer periods of protection rather than shorter; it would propertize more areas of information rather than fewer; it would impose a substantive set of standards of intellectual property on all states; and it would have few, if any, discretionary mechanisms allowing states to adjust levels of protection to suit their own interests. The reason Drahos believes the participants in the international original position would reject such a scheme is that they would be concerned at the potential impact of such a scheme on national sovereignty and the balance of power. We will explain this further.

Drahos argues earlier, in his discussion of Locke, that the power to adjust property rights within its territory is one of the most important powers a state has. It is 'a coping mechanism that states use to adjust to internal and external stresses' (Drahos, 1996, p. 188). From this perspective it is clear that a representative of a state in the international original position would not choose a global protectionist regime for intellectual property. Such a scheme allows economically powerful and information-rich states to set limits on the capacity of less fortunate sovereign states to adjust their property rights to suit their own circumstances.

Hegemonic states – which typically have a build-up of human capital at their disposal – could use a global protectionist scheme to gain permanent ascendancy over other states, thereby institutionalizing and legitimizing their control over the vital capital resource of information. Thus, a global protectionist regime 'presents any potential hegemon with a temptation of biblical proportions' (Drahos, 1996, p. 191). This hegemonic power would not simply be economic but, perhaps more significantly, would be cultural. This constitutes another objection for Drahos to a global protectionist regime. He worries that local or territorially based abstract objects may be appropriated for use in global markets, and that subsequently, economically powerful companies and states would become colonizers of other cultures. Sacred objects would become commercial objects, their meaning fading as the cultures they belonged to were swamped by trade.

We accept fully Drahos' economically based warnings over the threat which a global protectionist regime of intellectual property poses for the sovereignty of information-poor states. A strong regime of intellectual property rights simply does not suit non-developed states seeking to increase their base of human capital. It acts as a barrier to the diffusion of knowledge and the development of industry. A global protectionist regime can only suit economically powerful nations who possess the human capital, knowledge and research capacity that leads to the production of profitable abstract objects. However, Drahos' arguments concerning culture are less convincing. He blames intellectual property rights for the 'colonization' of less powerful cultures, and the transformation into commodities of previously meaningful abstract objects. But this process is a result of general trends concerning global economic inequality that allow members of one culture to travel the world experiencing other cultures as mere entertainment. Intellectual property rights have little to do with it. In fact, as we argue in the next chapter, the danger posed by tourism to the abstract objects of other cultures counts as a reason to *extend* intellectual property rights to minority cultural communities. In this way they would be able to regulate access to the cultural heritage that frames their lives, and to prevent the trivialization of their culture that Drahos is concerned about.

Drahos has one final warning concerning the threat posed by a global protectionist scheme: he says it could lead to the formation of global intellectual property factions. The specific danger Drahos identifies is global rent-seeking. Interests grouped around intellectual property might attempt to change the rules in order to enable them to engage in non-productive profit-seeking activities:

> Multinational elites might be tempted to increase their profits through the simple stratagem of persuading a supra-national body to ratchet up levels of protection for abstract objects already in existence. It is hard to see how the wealth transfers involved in such activities could be thought to be consistent with either Rawls' domestic principles of justice or those of the law of peoples. (Drahos, 1996, p. 192)

Drahos is clearly worried about the potential of powerful interests grouped around the intellectual property provision of the World Trade Organization seeking to extend levels of protection. It is argued that intellectual property only appeared as an item on the agenda of the Uruguay Round of GATT because of the influence of certain interests on the United States Government. Certainly, it had not previously been considered part of the purview of GATT. It was included at the insistence of the United States because of the strength of the cross-retaliatory measures available through GATT which were not available in the more traditional international forum for the regulation of intellectual property, the World Intellectual Property Organization. As Boyle writes, the USA's 'hard line over intellectual property is a relatively new phenomenon. In recent years, driven by some very effective lobbying and the spectre of the enormous losses to "piracy" that I quoted above, the United States government has dramatically changed its approach' (Boyle, 1996, p. 122). One could say, then, that a global protectionist regime only exists because of just the sort of factionalism Drahos warns about.

The Vicious Circle of Intellectual Property

Drahos argues that there is an inevitability about these threats, given the dynamic nature of intellectual property. He claims that as intellectual property is inaugurated in a society and gains a foothold in people's consciousness, so pressures for its expansion issue from those whom it has benefited. Those who benefit from intellectual property are interested in its expansion, not out of some altruistic desire to benefit society as a whole but to further their own interests. This is where the dangers of intellectual property come from; the interests of businesses favoured by intellectual property do not necessarily coincide with those of the public.

Drahos first describes this scenario in his chapter on Hegel. In Hegel's thought, the 'political state' is the population of a country as it is permeated with *Sittlichkeit*, or ethical life (Hegel, 1967). A state must be an ethical unit; without *Sittlichkeit* a state is corrupt. The state protects property rights as the basis of each citizen's interaction with each other. However, property is not unambiguously a good thing: certain dangers are inherent in it. In particular, it has the capacity to encourage subjective, selfish impulses in civil society that can threaten *Sittlichkeit* itself. The state can all too easily become the 'compliant arm' of civil society – a calamitous outcome if you regard the state as an essentially ethical institution based on the ideal of public service.

Intellectual property, believes Drahos, is, from such a perspective, especially dangerous, because it increases the pressure from civil society on the state as people realize the strategic business advantages of intellectual property rights. This process creates a 'feedback loop' of ever-increasing inequalities of information. Intellectual property is potentially worse than other forms of property because it 'represents an extension of property rights to an almost indefinite range of objects – scientific ideas, art, the genetic codes of nature – all of which fall within intellectual property's ever expanding domain' (Drahos, 1996, p. 87).

Drahos talks of property having four functions within a society: an appropriation function, an adjustive function, a self-defence function, and most importantly from our point of view, a planning function. Property rights are vital to rational actors, says Drahos, because they give them the security that allows them to plan. He asks two questions: first, what is the nature of planning in modern markets? And second, how might rational actors use intellectual property rights in order to help meet their planning needs? Drahos enlists Adam Smith's aid to address the first question. Smith (1976) famously argued that business people could not meet together even for recreational purposes without being tempted to conspire for their corporate ends against the public good. For Drahos there is no doubt that businesses will use intellectual property rights against the public interest if it is in their interest to do so.

The second question is rather more complicated. To answer it, Drahos makes use of the idea of intellectual property rights as a bulwark against the uncertainty of generating new knowledge, i.e., conducting research and development. There are three sources of uncertainty connected with research and development: nature itself, which sometimes means research simply fails to come off; the complexity of social systems, which may render present research obsolete when it eventually leads to a product; and the presence of competitors, which, other things being equal, acts as a disincentive to investing in research and development (R&D). Intellectual property rights address this last type of uncertainty by prohibiting free-riding on products and processes that arise out of R&D.

However, when intellectual property rights are instituted, those who benefit from them soon begin to realize that they could benefit more from their extension. If R&D is protected through intellectual property rights, the firm doing the research will get the pay-off from the R&D, while the free-rider gets nothing. Furthermore, the size of the pay-off accruing to the firm that does R&D will be greater, the stronger the intellectual property right they have. Clearly, therefore, it is in the interests of firms that invest heavily in R&D – generally large, capital-rich transnationals – to have the strongest intellectual property rights protection possible.

Intellectual property rights, then, are created to stimulate inventiveness, but they also have the effect of stimulating actors' interest in the property rights themselves. Firms know they can do better if they can get the protection afforded by intellectual property rights increased further. They are pursuing two strategies simultaneously: what might be called a strategy and a meta-strategy. In the language of game theory, the strategy is to pursue the best action within the game as it is presently constructed, while the 'meta-strategy' is to try to change the rules of the game in their favour.

But why does Drahos think that this behaviour is threatening to the liberty of others? To answer this question we need to turn to the idea of prevention in Drahos' thinking. Intellectual property rights crucially have a preventive function; that is, they prevent others from pursuing particular strategies in the market place. To use Drahos' own example, if a farmer grows cherries, then others are at liberty to imitate him and enter into competition with him. However, if the same farmer grows a particular, genetically-engineered type of cherry, which he owns the patent on, others cannot imitate him. One cannot claim that he is therefore immune from competition – other farmers can obviously grow other kinds of cherries to compete with him. Nevertheless, his patent acts as a constraint on the competitive behaviour of others. It also gives him an advantage if his cherry is superior – as it surely would be if he was bothering to grow it at all. The point is that intellectual property rights have anti-competitive effects; and if the justification of intellectual property rights is that they stimulate innovation and competition, then their anti-competitive effects must be taken seriously. Drahos writes that '[c]ompetitive markets are centrally to do with encouraging traders to imitate one another, while intellectual property rights are centrally concerned with preventing imitation' (Drahos, 1996, p. 135).

We have already given an indication of Drahos' belief that intellectual property rights stimulate interest in their expansion. Drahos now redefines this process in terms of traders in the economy pursuing 'preventive strategies', aimed at reducing competition against them. The most significant strategy, from our point of view, is the 'rights-expanding' strategy. This is the attempt to increase the scope, level and length of protection for abstract objects. Cornish analyses the process of rights expansion in intellectual property and sees two kinds of expansion: accretive and novel. The accretive process, the more significant of the two, is simply the expansion into new areas of intellectual property rights that were not originally intended for that purpose. The recent expansion of patents to include plant material is a good example of this process. The novel process is the institution of new forms of intellectual property right, such as those for semiconductor chips (Cornish, 1993, p. 55).

We would add that there is a third mode of rights-expansion, one that is even more significant for our purposes. That is the expansion of the 'breadth' of the permissible claim on a particular abstract object. Essentially, this means defining the abstract object to be protected as widely as possible in order to benefit the claimant. A classic recent example would be the granting of patents to Agracetus on all types of genetically engineered cotton and soybeans. As Drahos points out, the broader the patent, the fewer the opportunities for traders to compete: '...as the scope of the abstract object expands, it sets limits on the substitution possibilities that competitors can offer in the market-place. In other words, the degree of permissible imitation shrinks' (Drahos, 1996, p. 136). If the patent granted to Agracetus had been for a particular method of generating genetically engineered cotton or soybeans, or a particular trait, competitors would have been able to develop their own genetically engineered cotton in order to compete. Now, however, they must pay royalties to Agracetus (owned by Monsanto) if they wish to do so – which seems to defeat the purpose of competition. Here is a clear example of the negative effect intellectual property rights can have on economic liberty; companies are now unable to produce any genetically engineered products of two entire species because these are covered by intellectual property rights.

According to Drahos, these potentially harmful effects of intellectual property on negative liberty are compounded by the fact that self-interested actors tend to form factions around their property interests. These factions are more powerful than the individual members and push for changes in law that favour their members. In other words, the 'planning against markets' that Drahos identifies in the case of intellectual property is likely to be carried out by factions. Intellectual property owners are likely to organize into factions since '[f]actions form naturally around property because it sustains their power and way of life' (Drahos, 1996, p. 139). Factionalism in intellectual property is more likely, and more dangerous, than factionalism in other forms of property because of the considerable power of intellectual property rights to set constraints on the behaviour of others within the market. All property obviously has a certain constraint-setting effect, but because intellectual property rights relate to abstract objects with no natural boundaries the effect is much greater. Drahos concludes with a warning:

> These [preventive] strategies, we have suggested, operate against markets. They are part of the cost of intellectual property rights. It is a cost the dimensions of which have yet to be fully realized. And it is a cost which will continue to grow as more and more of economic life becomes entangled in privileges over abstract objects. (Drahos, 1996, p. 139)

Proprietarianism, the Authorship Myth and the Dangers of Intellectual Property

We have spent the last three sections elucidating Drahos' warnings about the threat that intellectual property poses to liberty. The central focus of this chapter, however, is *proprietarianism* in intellectual property. So how do proprietarianism and Boyle's 'author-vision' fit in with the dire warnings Drahos issues about intellectual property?

Boyle's message concerning the authorship myth and Drahos' message concerning proprietarianism and intellectual property are quite simple. Boyle's view is that the intellectual property system we have, because it discriminates against the protection of certain kinds of information, is unfair and unwise, given that it could lead to the public domain becoming a 'fallow landscape of private plots' (Boyle, 1996, p. 38). As we have seen, Drahos also believes present ways of thinking in the intellectual property system may diminish the intellectual commons. Additionally, in Drahos' view, intellectual property has the potential to damage the individual liberty of citizens and to create injustice. These outcomes can be avoided, but the proprietarian attitude towards intellectual property will, in Drahos' view, exacerbate them. This is because it fosters an uncritical attitude to intellectual property rights which blinds its adherents to the potential harms – indeed to the very idea that there might be harms – caused by intellectual property.

One of the ways proprietarianism fosters this attitude is by presenting intellectual property not as a privilege but as a right. This is an ideological process: as intellectual property becomes an accepted part of commercial life, it comes to be seen as an entitlement that must be respected, rather than as a privilege granted by it for the good of all. So property rights favour certain economic interests in society, who grow richer and more powerful as a result of them, and then use their position in society to spread beliefs concerning the arrangements surrounding property that made them rich. This sounds conspiratorial but it does not necessarily have to be; those involved in the commercial production of ideas will inevitably dominate the discourse surrounding it, and their terms are likely to be accepted as the standard ones. It is clearly in their interests that intellectual property is seen as 'right' or inviolable – indeed, they probably believe themselves that it is – so they are likely to refer to them in this way.

In the case of intellectual property this process is particularly dangerous, according to Drahos at least, because it presents privileges which limit the liberties of others in extensive ways as subjective or natural rights: 'Their distinct character and the threats they pose are clouded by a rhetoric of private property in which a universal subjective will is mobilized to defend the special interests of privilege seekers' (Drahos, 1996, p. 217). The costs to non-proprietors, and to society as a whole, of these privileges, are not taken into account.

Potential Solutions

Boyle and Drahos are, however, not simply concerned with criticizing the intellectual property system, but want to suggest how it can be changed for the better. The suggested solutions of each writer reflect their respective critiques. Drahos believes the problem with the present system is that a particular attitude to intellectual property has come to dominate thinking within the system. He therefore advocates the adoption of a different attitude: one that looks on intellectual property instrumentally rather than as a fundamental legal right. Boyle, conversely, sees the problem in terms of certain biases that lie behind the whole system as we know it. His preferred solution is to inaugurate new forms of intellectual property that do not rely on the author-vision but are capable of recognizing non-authored information. Boyle's solution is more practical than Drahos'; advocating a change in attitude is all very well but may not accomplish anything, at any rate in the short run. Boyle has demonstrated his determination to change the system by helping to draft a declaration on 'cultural agency/cultural authority', the Bellagio Declaration, signed by various NGOs and activist groups around the world (Boyle, 1996, pp. 192–200). As he says, his 'goal is prescriptive as well as descriptive' (Boyle, 1996, p. 165). The solutions promoted by the two writers, are not, however, incompatible. Boyle's new forms of intellectual property assume the attitude of instrumentalism that Drahos advocates. In what follows, Drahos' reasons for advocating instrumentalism are discussed, followed by an evaluation of Boyles' more concrete suggestions.

By 'instrumental', Drahos means that intellectual property rights should be used to serve our goals and ought therefore to be flexible and suited to present circumstances. We can identify three ways in which Drahos attempts to persuade the reader that an instrumentalist attitude is preferable. The first is directly consequentialist: Drahos simply presents what he believes to be the potential consequences of an uncritical attitude to intellectual property. We have already considered Drahos' views on the consequences of intellectual property, so we do not need to say any more on that subject.

The second method of persuasion Drahos uses is Rawlsian: in endorsing Rawls, he is embracing Rawls' instrumentalism of property. In Rawls' theory, property rights are not primary moral rights. People's property must be respected, and as Drahos points out, certain instrumental uses of property are forbidden, but the precise form of property rights is flexible and can be altered. Rawls does not argue this point at any great length, but his views are clear in certain passages, such as that concerning competition policy and economic efficiency, in which he claims that in the interests of such goals, 'the scope and definition of property rights may be revised' (Rawls, 1971, p. 276). Property rights in Rawls then, do not have the 'iron-clad guarantees of safety' that they have in entitlement theories such as that of Robert Nozick (1974) (Drahos, 1996, p. 178). They serve the goal of justice rather than forming a part of justice. Drahos does not attempt to justify Rawls' instrumentalism; he merely adopts it.

The third method used by Drahos to persuade the reader to adopt an instrumental attitude to intellectual property is historical; he demonstrates that the original intention of most intellectual property forms was not to respect or protect natural entitlements but to serve the public good. We will, following Drahos, give a brief account of the origins of the patent system, as this is the most important intellectual property form from our point of view.

Although the first patents were granted in Venice, it was in England that patents were inaugurated in their present form. Drahos presents the historical record to show that English law has always been highly instrumental in its treatment of patents. This is most manifestly shown in the case *Darcy* v. *Allen* of 1602, the case that laid the foundations for English legal thinking on patents. At the time, King James I was in the habit of selling monopolies to raise revenue. At the turn of the seventeenth century he granted a patent (monopoly) on the making of playing cards to a Mr Darcy. However, the patent was disallowed in the courts on the grounds that it was a profound interference in the liberty to trade, and it was therefore void in common law. The challenger Mr Allen's counsel, however, argued that there was one exception to the rule that monopolies were forbidden. In cases where a useful trade had been brought into the country by a person, 'the King may grant to him a monopoly patent for some reasonable time, until the subjects may learn the same, in consideration of the good that he doth bring by his invention to the commonwealth: otherwise not' (quoted in Drahos, 1996, p. 31).

Patents were allowable, then, only if they resulted in the raising of living standards via the introduction of some new technology into England. This same attitude was displayed in the Statute of Monopolies of 1623, which, although it declared all monopolies to be invalid because they did not lead to the 'publique good', made an exception in the case of monopolies granted to the inventors of new methods of manufacturing. However, as Drahos points out, 'The Statute made clear that patents belong to inventors by virtue of a privilege and not a natural right of some kind' (Drahos, 1996, p. 32).

Drahos also shows that trademarks have similarly instrumentalist origins. It is clear that they were originally intended to serve the public, distinguishing one manufacturer's goods from another. In economistic jargon, they have a 'general communication function', transmitting information to consumers regarding the origins and properties of goods. They are valuable to consumers because they lower the 'search costs' of the particular goods they are looking for (Drahos, 1996, p. 205). However, Drahos believes the proprietarian tide has shifted the perception of the dominant purpose of trademark law to protecting the interests of traders. The language of property and rights has replaced that of monopoly and privilege, and, in an example of Cornish's accretive process, colour, sound, smell and taste are now protectable under trademark legislation, where before only signs were protectable. Furthermore, trademarks are often now tradable entities; this is in direct contravention of the original purpose of trademarks and could even be said to be deceptive towards the public, who can no longer be sure if goods carrying a certain mark are indeed the goods of the particular manufacturer they seek (Drahos, 1996, p. 206).

In his writing, Drahos wishes, therefore, to turn back the tide to the instrumentalist origins of intellectual property rights, in the hope that this would shift the emphasis away from the protection of commercial interests towards the protection of the public interest. In contrast to Drahos' attitudinal approach, the final chapter of Boyle's book *Shamans, Software and Spleens* is devoted to more concrete proposals. From our point of view the most important of these suggestions is Boyle's suggestions for a *sui generis* system that could incorporate intellectual property rights for communities in traditional varieties and botanical knowledge. He quotes the Bellagio Declaration on Cultural Agency/Cultural Authority, which he himself helped to draft:

> We favor a move away from the author vision in two directions; first towards recognition of a limited number of new protections for cultural heritage, folkloric productions, and biological 'know-how'. Second, and in general, we favor an increased recognition and protection of the public domain by means of expansive 'fair use protections', compulsory licensing, and narrower initial coverage of property rights in the first place. (Boyle, 1996, p. 169)

Boyle recognizes the irony that, after his long critique of intellectual property rights and his complaint that there are too many such rights, his solution should be to inaugurate yet more new forms. However, his critique is not of the very idea of intellectual property rights, but of the author-centred conception of intellectual property rights that, he believes, dominates the intellectual property system and produces adverse effects on the public domain and, in turn, on the production of information itself: 'My point is not that we need fewer intellectual property rights, or that we always need more intellectual property rights. Rather, my point is that an author-centered system has multiple blindnesses and that we should strive to rectify some of them' (Boyle, 1996, p169). Similarly, Drahos' extensive critique of intellectual property rights was not intended as an argument for their abolition, but for a replacement of the dominant proprietarian conceptualization of them by an instrumentalist conceptualization designed to protect the public good.

CONCLUSION

A cursory glance at the present politics of genetic resource control reveals that the primary subject of controversy is proprietarian intellectual property rights. It is the extension of proprietarian intellectual property rights to genetic material that is responsible for the political controversy that now surrounds genetic resources. In a classic example of the power of a global hegemon, US-style strong proprietarian intellectual property rights have now been extended worldwide via the machinery of the World Trade Organization. The contribution of Boyle and Drahos is to point out the dangers of this process. As they make clear, the proprietarian intellectual property right system of the western world – now, by default, the system of the entire globe – has certain critical shortcomings that are very serious. It ignores the potentially damaging effects of intellectual property on individual liberty, on international justice and the egalitarian democratic ideal itself. Furthermore, it rewards the genetic innovations and discoveries of western science with stringent intellectual property rights, while treating other kinds of genetic information as raw material, thus increasingly the likelihood that these sources of information will die out.

One of Boyle's solutions aimed at rectifying the imbalances of the proprietarian intellectual property system is a *sui generis* system for the protection of certain kinds of non-authorial information concerning genetic resources. This includes the genetic information contained in traditional varieties of crops, and the information concerning the properties of plants that is held by indigenous and tribal peoples the world over. In the next chapter, we address the question of an intellectual property regime for such information, and develop an argument for community intellectual property rights based on autonomy.

NOTES

1. 'Seed-saving' is the practice of saving seed from one harvest to plant for the next. It is true that elite varieties of some crops, as we saw in Chapter 2, are hybrids and therefore lose their vigour in the second generation; seed cannot be saved from such varieties. But some crops' elite varieties are not hybrids and seed can still be saved.
2. The argument that human freedom is illegitimately curtailed by patenting could be framed in either categorical terms or consequentialist terms. Nevertheless, even the categorical formulation of such an argument would presuppose an instrumentalism of intellectual property.
3. In the USA, only publication in the USA is taken as evidence that an invention is not new. This, it seems, is how W.R. Grace were able to patent 'inventions' taken from the seed of the neem tree. Despite the fact that the properties of neem have been common knowledge in India for many centuries, Americans were formally ignorant of these properties and so the patent (which applies worldwide, including in India, as a result of GATT) stands.
4. We are here employing the terms used by Drahos, who in turn follows Tully (1980). The same concepts are sometimes referred to by the terms *res communis* (positive community) and *res nullius* (negative community).
5. Raz, *The Morality of Freedom*; Kymlicka, *Liberalism, Community and Culture*.
6. Svatos points out that 'copycat' inventions are common in the case of pharmaceuticals. Drug manufacturers can modify proprietary pharmaceuticals, producing essentially the same drug but without infringing the patent (Svatos, 1996, p. 121).
7. Rawls has two principles of justice. The first states that every person is to have an equal right to the most extensive basic liberty consistent

with a similar liberty for others; the second claims that social and economic inequalities should be arranged so as to (a) maximize the position of the least well-off, and (b) be attached to positions open to all (Rawls, 1971, p. 60, p. 83).

8. Any gains in material well-being, even of the least well-off, are forbidden if they interfere with political liberties. In this case, then, property rights in information are disallowed if freedom of information is subverted. This would seem to rule out all intellectual property rights because they all present limits on access to certain information and its use.

4 Community Intellectual Property Rights

INTRODUCTION

The controversy over genetic resource control stems, as we argued in Chapter 2, from the extension of intellectual property rights to plant and other genetic material in the industrialized world. Third World dissatisfaction over intellectual property rights has manifested itself via a number of responses. The three most significant of these responses are (i) the demand for some sort of intellectual property rights for indigenous and farming communities in their traditional plant varieties,[1] farming practices and botanical knowledge (we refer to these rights from now on as 'community intellectual property rights' or 'community IPRs'[2]); (ii) the move to assert national sovereignty over genetic resources; and (iii) the campaign for a distribution based on the common heritage ethic. This chapter is devoted to the first of these responses and it has three sections. In the first section we explain the importance of traditional varieties and knowledge, and how this importance has come to be recognized by the industrialized world in the past decade or so, and go on to give a brief account of the rise of the 'indigenous peoples question' in international politics and to show how this relates to the demand for community intellectual property rights.

In the second and third sections, we consider two justifications of community intellectual property rights. In our view, it has never been adequately explained *why* indigenous and farming communities should be granted rights over their traditional varieties and knowledge. This is clearly something that needs to be done if international and domestic policy-makers are to be convinced of the desirability of community IPRs. At the time of writing, despite some practical moves in some areas towards something akin to the extension of a form of intellectual property rights to communities, there is still no

international or national *sui generis* system for indigenous varieties and knowledge. Discussions at an international level
reveal a considerable amount of scepticism on the part of some
industrialized countries towards community IPRs. This signals
that the argument for community IPRs has not been won –
possibly because an adequate argument for them has never
been made. We attempt just such an argument here. In Section
II, we discuss and reject the theory that communities are
entitled, by virtue of labour, to intellectual property rights. In
Section III, we develop an argument in favour of community
IPRs based on the value of individual autonomy; we argue, following Kymlicka, that for autonomy to flourish in individuals, a
strong cultural structure is necessary. Special cultural rights are
needed in order to protect cultures, and intellectual property
rights in genetic resources and related knowledge are examples
of such rights. We also discuss what sort of rights are justified
from the point of view of the theory we develop.

SECTION I: THE MEANING AND SIGNIFICANCE OF COMMUNITY INTELLECTUAL PROPERTY RIGHTS

The Importance of Traditional Varieties and Knowledge

Since the time of early colonialism the North has recognized
the value of the South's genetic resources; this has included the
indigenous peoples' traditional crop varieties as much as the
tropics' wild germplasm. The germplasm from both sources has
contributed, and continues to contribute, a colossal amount to
the economies of the North, and is partly responsible for the
structure of the modern global economy and the developing
world's disadvantaged place within it. Yet, as should be clear by
now, the indigenous and rural communities of the developing
world have never been compensated for the removal of this
germplasm. Until recently it was seen as the common heritage
of mankind, owned by no individual, community or country and
freely available to all.

If the value of the germplasm belonging to the indigenous
communities of the world has always been recognized, the
knowledge possessed by members of these communities with
regard to the biological diversity that surrounds them has been

largely ignored until very recently. The assumption of western cultural superiority, particularly in its science and technology, has meant that the possibility that indigenous knowledge might be useful to the industrialized world has rarely been considered. Indigenous knowledge systems have been thought of as 'unscientific', 'backward', or 'primitive'. Only in the last decade, or perhaps two decades, has this perception begun to change, as the work of applied anthropologists has revealed, on the one hand, the epistemological complexity and sophistication of indigenous knowledge systems, and on the other hand, the potential usefulness of indigenous knowledge to the industrialized world.

Indeed, the last few years have seen something of a clamour for indigenous knowledge, as it has finally been recognized that indigenous communities often hold the key to the value to humanity of the world's biodiversity. 'Ethnobotany' has become a boom subject in both academia and business in the last decade, as the value of indigenous peoples' plant-related knowledge has been recognized. This knowledge is extremely important for the future of humankind, and is valuable financially to Northern companies who hope to use it to develop lucrative – generally medical – products. Almost all the major pharmaceutical companies of the world are involved in some way with ethnobotany, while specialist ethnobotanical institutes and companies have been set up to catalogue this knowledge. This clamour has given rise to worries over the effects of such 'knowledge prospecting' on indigenous communities, and also the fairness of using such knowledge to develop industrial products that have no connection to the original cultural context from which the knowledge came. The call for community intellectual property rights is partly a response to these worries.

Many examples can be given to demonstrate the importance and value of traditional varieties and knowledge. Most of these examples are taken from a list compiled by the Rural Advancement Foundation International (RAFI), a nongovernmental organization (NGO) based in North America and concerned with issues of genetic resource control. Examples of traditional varieties that have come from Third World farming communities and have contributed to the agricultural economy of the industrialized world include the Ethiopian barley variety, worth $150 million in the United States every year; the Turkish wheat, a sample of which was taken from Turkey and is now worth $50 million in the USA Northeast; and the West African maize containing the only genetic resistance to Southern Corn Leaf Blight ever found, which was used in breeding resistance to the blight in the USA and eradicated a disease that cost US agriculture a billion dollars in the 1970s (RAFI, 1996a, pp. 57–81). In these cases, it is not the traditional variety *per se* that is grown and that creates the extra revenue for Northern agriculture, but specific genes that have been bred into existing 'elite' lines to offer new characteristics such as disease resistance or increased yield. Nevertheless, the farmers who cultivate the traditional varieties from which such characteristic come are fully aware of the usefulness of the characteristics; indeed, these characteristics are the reason why such varieties are still grown by them.

The above examples are instances of cultivated traditional crop varieties contributing to Northern agriculture. In addition, there are numerous cases of traditional knowledge leading to medical discoveries and other advances. RAFI report a Peruvian medicinal tree whose powers were known to local communities and which is now being used to manufacture Stimulon, an anti-AIDS drug. There is also the case of the Kani tribe of India leading researchers to the *jeevani* plant, which the tribe has been using for centuries and whose berries have ginseng-like energy-giving qualities (Jayaraman, 1996). Another example from the RAFI list is the tikluba plant, used as an anti-coagulant by the Ure-eu Wau Wau tribe of Amazonia and now, thanks to the assistance of the tribe, being developed into a commercial product by Merck Inc. (RAFI, 1996a). These are just a few examples; the knowledge of indigenous communities is now highly prized by western firms and they are doing all they can to get their hands on it.

The Demand for Community IPRs

The idea that one way of protecting indigenous and farming communities from outside interests seeking to profit from their knowledge is to inaugurate some *sui generis* form of intellectual property rights originated in the mid-1980s. Michael Huft describes how the idea was first mooted amongst agricultural scientists, anthropologists and 'cultural advocates' in the developed world, and that '[o]nly in the last five years have the twin ideas of compensation and intellectual property rights begun to be discussed by people in developing countries, in particular among government officials' (Huft, 1995, pp. 1680, 1685). In 1996 Tom Greaves wrote that 'IPR for indigenous peoples has become a "hot-button"' issue' (Greaves 1996, p. ix).

The campaign for community IPRs is an outgrowth of the wider campaign for indigenous peoples' rights in general.[3] The position of indigenous peoples has become an increasingly salient political issue over the last few decades. It has been recognized that indigenous peoples have a unique experience and position in the world and that their needs are correspondingly unique. The policy of assimilation, whereby it was thought that the best way to protect indigenous peoples' rights was to treat them just like other members of the wider society, has now been discredited. By and large, it seems, indigenous peoples want to live in their traditional communities; they do not want their communities to die as a result of their special interests being ignored.

Since the 1970s, indigenous groups have been working in the international arena to ensure that their voice is heard. In 1977 their efforts gave rise to the International Non-Governmental Organization Conference on Discrimination against Indigenous Peoples in the Americas, held in Geneva. This was attended by indigenous peoples' representatives from the western hemisphere and, in James Anaya's words, 'contributed to forging a transnational indigenous identity that subsequently expanded to embrace indigenous peoples from other parts of the world' (Anaya, 1996, p. 46). Since then representatives of indigenous peoples have been well organized and their presence in international fora is now taken for granted. An example of this is that indigenous peoples, as a group, have a guaranteed post on the Secretariat of the Convention on Biological Diversity. And as we shall see in Chapter 5, on national sovereignty over genetic resources, indigenous peoples are the subject of a series of international agreements, some of which grant them a measure of control over the natural resources within their territory.

The idea that indigenous peoples should have control, as a community, over their natural resources is therefore a familiar one. Nevertheless, no agreements on a community IPR arrangement have yet been made, although they are being seriously discussed in a number of fora, most notably the FAO's conferences on Plant Genetic Resources for Food and Agriculture and the Conference of the Parties to the Convention on Biological Diversity. Both the International Undertaking on Plant Genetic Resources and the Biodiversity Convention contain clauses that could act as the basis for a *sui generis* community IPR system (Correa, 1995, p. 73). Article 8 (j) of the Convention, while more vague than the Undertaking in its application to community IPRs, recognizes that the respect, maintenance and preservation of the 'knowledge, innovations and practices of indigenous and local communities embodying traditional lifestyles [is] relevant for the conservation and sustainable use of biological diversity' (Grubb et al., 1993, p. 78). National legislation should promote the application of this general principle 'with the approval and involvement of the holders of such knowledge, innovations and practices' (da Costa e Silva, 1995, p. 546). da Costa e Silva points out that 'when the Convention discusses knowledge, innovations and practices and entitles local and indigenous communities to be their *holders*, it links these concepts with the vocabulary typically used for the definition of the proprietor of an intellectual property right' (da Costa e Silva, 1995, p. 546). Discussions from the Conferences of the Parties to the Convention confirm that community intellectual property rights are firmly on the agenda.

In addition to the Undertaking and the Convention, one of the most significant agreements for our purposes is the World Intellectual Property Organization/UN Economic, Social and Cultural Organization Model Law on Folklore, which was approved in 1985. The Model Law has not been formulated into a legally binding Convention, and excludes scientific innovation, which means that plant varieties and botanical knowledge are not included, but it nevertheless acts as a precedent for the principle that indigenous communities can own their own cultural forms and creations. According to RAFI, it has three unique elements – in terms of IP law – that are particularly relevant to any future protection of biological products processes: first, it recognizes that 'communities' can be legally registered owners of intellectual property; second, it states that community innovations are not necessarily fixed and finalized but can be ongoing and evolutionary and still be protected; and third, it asserts that any rights granted to communities under the

Model Law would not be time-limited, unlike standard forms of intellectual property (RAFI, 1996b, pp. 46–7).

There are also precedents in national law for community IPRs. The Brazilian Bill 2057/91, concerned with updating indigenous rights in accordance with the 1988 Brazilian Constitution, states that indigenous communities, societies or organizations have the right to apply, directly or indirectly for a patent, utility model, industrial model or industrial design protection, for their indigenous knowledge or models. It also states that where indigenous knowledge has led to an invention, a community will be considered the co-proprietor of any resulting patent (da Costa e Silva, 1995, p. 549). However, it is unclear how far these provisions are in accordance with subsequent obligations arising out of the GATT agreement of 1994.

IPRs for communities have, then, gone beyond the stage of being a purely theoretical concern of activists and concerned individuals, and are firmly on the international political agenda. However, little progress has been made in reaching an agreement either on their desirability or on the form they might take. We believe this is partly due to the fact that a well argued and logically powerful case for community IPRs has never been made. In the next section we attempt to make such a case. We begin by discussing the application of entitlement theories to the issue of IPRs for communities.

SECTION II: THE ENTITLEMENT THEORY OF COMMUNITY INTELLECTUAL PROPERTY RIGHTS

This section examines and rejects the view that cultural communities are morally entitled to intellectual property rights in plant genetic resources and their associated knowledge. We take intellectual property rights to mean property in an intangible or abstract object, and here IPR refers to any legal recognition that a creation or resource is the intellectual property of a specified community. This community can be either an 'indigenous' or a 'non-indigenous' farming community.[4] It is important to note that community IPRs are distinct from 'Farmers' Rights' (Crucible Group, 1994, p. 34ff; Brush, 1996, p. 139). Although community IPRs have been incorporated into some conceptions of 'Farmers' Rights',[5] we take the term 'Farmers'

Rights' to refer to a variety of arrangements, some of which remain mere suggestions, that would assist farming communities in conserving their agricultural biodiversity. The conception of Farmers' Rights adopted by the Food and Agriculture Organization would, first and foremost, oblige those using plant genetic resources from traditional crop varieties collected from farming communities, primarily in the developing world, to pay money into a central fund, which would then be used to set up *in situ* biodiversity conservation programs (FAO, 1989, p. 13, n.1). 'Farmers' Rights', then, as Stephen Brush points out, recognizes the ongoing contributions of cultural communities' farmers *without* directly attributing ownership to, or commercializing, their resources and knowledge (Brush, 1993, p. 662). By contrast, IPRs necessarily entail conceptions of ownership.

A moral entitlement theory of property is one that sees property as a primary moral right that is not justified by reference to consequences. Entitlement theories attribute rights in specific things to specific individuals or groups on the basis of a special relationship between object and claimant. Utilitarian theories that justify property by arguing that it promotes society-wide efficiency are not, therefore, moral entitlement theories, and neither are those who argue that property is necessary to ensure peoples' ability to express themselves. The labour theory of property, which claims that ownership in something arises as a result of working on that thing, is paradigmatically an entitlement theory: to have laboured on something is the special relationship that gives rise to the moral entitlement. Consequences are irrelevant.

As we saw in the previous section, there have been agreements made between indigenous communities and other bodies that could be seen as conferring a sort of intellectual property right on the community in question. Recently, for example, the Kani tribe of the Indian state of Kerala was granted a form of intellectual property right in the active ingredient of a plant-derived drug named *jeevani*, which is said to combat stress. The Kani received a $25 000 'know-how' fee and will also gain a share of the 2 per cent royalty on any future sales of a medicine developed from the plant (Jayaraman, 1996). These intellectual property rights do not fall into any of the standard western categories – patents, copyrights, trademarks and trade secrets – but they can be seen as an implicit form of intellectual property right because they recognize a specified community's ownership rights in a specified substance. Nevertheless, such ad hoc contractual arrangements, while a good sign, are unsatisfactory as compared with a *sui generis* intellectual property system, as we shall argue in the conclusion to this chapter.

There are many consequentialist arguments in favour of granting community IPRs to communities. They might be necessary to conserve biodiversity (Margulies, 1993), for example, or to protect traditional ways of life (Greaves, 1995, p. 203). However, this section is not concerned with these consequential questions. It is intended as a rebuttal of only one argument: the idea that cultural communities hold moral entitlements to IPRs in plant genetic resources (PGRs). In the following section we argue for IPRs for communities on the basis of the value of individual autonomy.

In the previous chapter we saw that throughout history, the predominant norm with regard to access to genetic resources was one of 'common heritage'; this was taken to mean that wild resources and landraces were freely available to companies seeking to use traditional varieties and wild genetic resources to develop new plant strains and pharmaceuticals. Some commentators have argued that this treatment of 'raw' germplasm as an unowned resource, and therefore freely available, by the plant breeders of the 'North', was robbery, and that the subsequent protection of 'elite' varieties of plants by these breeders is rank hypocrisy, adding insult to injury:

> ... in manipulating life forms you do not start from nothing, but from other life forms *which belong to others* ... third world countries [are] the original owners of the germplasm ... wild material is 'owned'; by sovereign states and by local people.[6] (Shiva, 1991, pp. 2745, 2746, italics added)

National sovereignty is now, of course, an accepted norm in access to genetic resources; however, the idea that 'local people' own genetic resources – wild or cultivated – is not yet fully accepted. Writers like Shiva have argued that if, by dint of the work they have carried out in creating new strains of crops, First World innovators are entitled as of right to patents on their inventions, then so are Third World communities entitled as of right, for the same labour-based reasons, to some form of IPRs in their landraces and botanical knowledge.

This position takes for granted an entitlement theory of intellectual property. We argue, however, that from the point of view of such a theory, there is no hypocrisy involved in asserting IPRs for special genetic stocks, yet denying them for original germplasm and cultural knowledge; indeed, in our view, an entitlement theory cannot ground intellectual property rights for communities in any sort of genetic resources or traditional knowledge. If Third World campaigners wish to argue for intellectual property rights for cultural communities, they should abandon the idea that these communities are entitled *by right* to intellectual property over genetic resources, and instead concentrate on consequentialist arguments, such as the autonomy-based argument propounded in the following section of this book.

The remainder of this section is divided into two parts. The first part deals with the question of community ownership of traditional varieties or landraces; the second part deals with community ownership of cultural knowledge. This division is based on convenience; essentially the same arguments apply against the application of entitlement theory in both cases.

Entitlement Theory and Traditional Varieties

There are several ways in which intellectual property rights can be justified. The principal distinction between types of justification for intellectual property rights, as with other forms of property, is between consequentialist and natural rights-based theories. The justification usually invoked by economists for a system of intellectual property rights is utilitarian, which is a consequentialist theory. However, the alternative idea that creators are entitled by right to a form of intellectual property in their creations is also a widely held belief. Certainly, entitlement theories of this kind have much force in our society when applied to ordinary tangible forms of property. Edwin Hettinger is right to suggest that '[p]erhaps the most powerful intuition supporting property rights is that people are entitled to the fruits of their labour' (Hettinger, 1989, p. 36). By contrast, Cornish has argued that certain features of the patent system that limit the inventor's entitlement – the need to show that the innovation is novel and inventive, and that it is being adequately 'worked' in the territory that has granted that patent – suggest that 'the real justification for the system lies in its economic impact' (Cornish, 1993, p. 50, note 10). But there seems no reason to believe that there is a 'real' justification. There are simply some arguments for our legal arrangements that are satisfactory and some that are not. As Horacio Spector writes: 'Applying Locke's [labour-based entitlement] theory to intangible property does not appear to be far-fetched' (Spector, 1989, p. 271). The limitations that Cornish talks about seem eminently compatible with an entitlement theory; in any case, the entitlement theory of property is a moral theory and is not hidebound by existing law. The mere fact that the goal of those who control the patent system – politicians and, primarily, judges – is avowedly utilitarian does not invalidate other moral theories seeking to argue for or against it; an entitlement theorist would argue that if certain features of patent law contradicted entitlement theory, then they ought to be done away with. Following Spector, therefore, we assume that there is no prima facie reason not to apply entitlement theory to intellectual property. The question of whether entitlement theory is a satisfactory justification for intellectual property *in general* is not one that concerns us here; we only wish to see whether it can be sensibly applied to the case of intellectual property rights for communities.

The obvious place to start is with the labour theory of entitlement. On the labour theory of intellectual property (IP) ownership, it could be argued that agricultural communities own their landraces by virtue of the fact that they have created them through centuries of seed selection and innovation. In claiming landraces as common heritage, western governments and companies are ignoring this labour and the entitlements it creates. This position is summed up by Daniela Soleri et al.:

> To those supporting farmers' intellectual property rights in their folk varieties, the effort and knowledge of indigenous farmers involved in creating and maintaining folk varieties implies the need for recognition on an equal footing with that of plant breeders and molecular biologists. They see the communal effort in developing folk varieties as an integrated part of making a living over generations to be as legitimate as the individual efforts of scientists in formal, segregated work settings in the laboratory or field plot. (Soleri et al., 1996, p. 24)

The belief that the work of generations of a community's farmers is as entitled to protection as is the work of modern scientists is also voiced by Vandana Shiva:

> Patenting gives monopoly rights on life forms to those who manipulate genes with new technologies, totally disregarding the intellectual contribution of generations of Third World farmers, who for over 10,000 years have experimented in conserving, breeding and domesticating plant and animal genetic resources. (Shiva, 1990, p. 46)

However, despite its apparent plausibility, the argument that a community's labour can create entitlements to property that adhere to that community, cannot be sustained. It raises several issues which can best be brought to the fore with a brief account of the labour theory of ownership.

The seminal account of the labour theory of ownership is, of course, that of John Locke (1988). In Locke's theory, the world's resources are *res nullius*: owned by no one and available for appropriation by anyone. But when someone mixes their labour with a natural resource, they come to own that resource. This is because people own themselves[7] and by extension, their labour. They are therefore entitled to the product of that labour. In answer to the question of why mixing one's labour is a way of coming to own what you do not own rather than losing something you do own (Nozick, 1974, pp. 174–5), Locke replies that labouring on something adds value.

The Lockean theory of ownership has been criticized on many grounds. For example, it has been argued that labour does not always create value; that the labour criterion is too vague; that the self-ownership thesis is atomistic and ethnocentric and fails to recognize that labour is a social activity; and that the labour theory cannot provide entitlements to *intellectual* property. We do not wish to enter those debates here; we simply want to see whether, even if the labour theory of intellectual property is sound, it can be consistently applied to the case of communities, IPRs and landraces. In a labour theory of intellectual property, it is the performance of a certain act – labouring – that creates an entitlement on the part of the labourer to certain rights concerning the product. The crucial questions here, then, are first, whether communities can be seen as actors capable of performing entitlement-creating acts, and second, whether the development of landraces qualifies as such an act. We answer 'no' to both questions, which are dealt with here in turn.

Are Communities Capable of Labouring?

There is a well-known tradition in political and social philosophy that sees communities, or nations, as independently existing organic entities that have 'lives' of their own, above and beyond the lives of their individual members. Clearly, communities do not have easily identifiable personalities like individuals, but some writers and traditions, from Rousseau (1973) to the romantics to Hegel (1967) and to the fascists, have pointed to a certain 'guiding consciousness', or the 'general will', in order to indicate that communities are greater than the sum of their individual parts. More recently, writers such as Vernon Van Dyke and Lawrence Johnson have argued that communities often cannot be accounted for as an aggregate of individuals and that they have a moral standing of their own, apart from the individuals who make them up (Van Dyke 1985; Johnson 1991).

However, even if we accept the controversial proposition that communities are independent entities that have moral standing, we do not have to accept that a community is capable of performing entitlement-creating acts – labouring – and thereby coming to own property. For entities to perform labour, they need to be in possession of many mental tools: a conception of the future, an idea of their own good and of how to achieve that good, the knowledge of how to perform the act that will achieve their aim, and so on. In short, to labour, an entity has to be capable of executing a rational plan aimed at some end. It is hard to see how a community, even if it were an independently existing entity, could achieve such a thing without a brain of its own. Unless we embrace organicism, a community, then, just does not seem to be the kind of thing that can engage in labour. It is individuals and individuals only who engage in labour (whether by themselves or collectively), and, therefore, on the labour theory of ownership, only individuals can create entitlements for themselves and come to own property.

However, let us for the moment embrace organicism and assume that communities can perform acts of labour. Perhaps we can postulate that when an individual engages in labour, the community is in fact working *through* that individual in order to fulfil a community good that cannot be reduced to individual goods. This might give us a reason to confer rights on communities. But there is a serious practical objection to conferring such rights on the basis of an organicist theory. This is that any decision over who can use the germplasm in plant breeding and biotechnological research and production becomes impossible to make. In order for communities to be able to allow their landraces to be used for research, we would have to postulate that these communities, as singular organic entities, had a Rousseauian 'general will' that could be discovered. But how do we discover this general will so that landraces can be transferred – for payment and other benefits – from their original communities and used in modern plant breeding? Clearly, voting procedures currently conducted will not suffice because a vote is merely the aggregate opinion of the individuals who happen to be eligible to vote at the time of voting, rather than the community's enduring 'general will'. Communities exist over many generations – that is what communities are – and so a snapshot of the opinion of the current generation cannot automatically be assumed to reflect the will of previous and future generations. Since the will of the community, assuming it exists, cannot be discovered by any practicable means, any IPRs we grant over landraces would then, have to be, permanent, and the germplasm itself would be forever inalienable: permission for outsiders to use landraces could never be granted. An absurd situation, unwanted by both plant breeders and (most) traditional communities, but one we unavoidably arrive at if we take the community to be an organism.

The reader will recall the terminology used by Boyle in the previous chapter on the shortcomings of intellectual property. He argued that the present system of intellectual property was blind to certain forms of information production, including the production of traditional varieties by communities. This illuminates the present problem: communities cannot be seen as 'author-figures', and therefore cannot be identified as intellectual property holders under a system characterized by the myth of entitlement and the romantic author. This is, of course, unless we regard the community as a single 'creator' entity, but we have seen the difficulties of this idea.

The reader might object to our rejection of the idea that communities can create entitlements to property through labouring, on the grounds that there are many groups in society today that do own property as groups. The most obvious example is that of a joint stock company. There is no individual owner of such a company, yet the company, which is after all a group of people, certainly owns its capital. However, in this case it is clear that ownership can be reduced to individual ownership. Although the decisions in a joint stock company are taken by the directors and employees, ownership, and with it control in the final analysis, resides with the shareholders. The shareholders do not by any stretch of the imagination constitute an organic, collective entity; they are individuals who own shares of the company individually, because they have each bought equity in it. The shareholders' decisions at an AGM bind the company, but in a cultural community such snapshot decisions are not binding.

Perhaps a better example, and one that gets closer to the relevant issues, is an agricultural commune, where the wheat harvest is communally owned by the members and outsiders are not entitled to a share of it. Now we *could* account for this ownership by saying that the commune has laboured on the wheat, has brought it to harvest and is therefore sole owner of it. In this case it seems that the group has indeed created an entitlement for itself by virtue of its labour. However, we can more accurately characterize this situation by looking at it in individual terms. It is not that the commune, as a singular entity, has laboured on the wheat, and therefore owns it *qua* commune. It is rather that all the individual members of the commune are entitled to a fair share of the wheat because it is they who jointly created it. The reason why outsiders are not entitled to a share, on a labour-entitlement theory, is because they have not individually contributed.

Is the Creation of a Landrace an Entitlement-Generating Act?

One might ask why we cannot view landraces in the same way that we view the wheat in the above example: that is, why cannot we say that landraces are owned collectively by the community, seen as a collection of individuals, rather than as an organic entity? We might justify this by arguing that the individual members of the community, who have been responsible for the creation of the landrace, have alienated their individually acquired rights to their community, in the same way that scientists employed by the plant breeding companies of the North alienate their rights over their creations to the company (in return for a steady wage and the materials with which to create new strains). However, in the case of both the commune and its wheat, and the scientists who create new crop strains, the point is that a group of people come together at the same time to collaborate, to work towards a common goal (creating a product). To use Boyle's terminology, there is an identifiable author-figure – which in this case happens to be a group of 'authors'. Through that common labouring the authors come to own shares in the product. Under the labour theory, those who made the effort own the product of it – unless a previous contractual arrangement has been made whereby the product is signed over to another body, such as the company that employs the scientists who developed the new strain.

Now, it is true that cultural community innovation is ongoing, and individual farmers continually experiment with variations of the landraces they use, often in a systematic and rigorous fashion (Hobbelink, 1989, pp. 135–47; De Beof et al., 1993; Fowler, 1995, pp. 221–2). We are prepared to entertain the idea that this innovation could qualify as entitlement-creating. However, if it does, the entitlement would accrue to the individual farmer(s) and not the community. If the experimentation was collaborative, in an entitlement theory only the individuals who collaborated would be entitled to IPRs in the product. Moreover, the IPR would only apply to that particular new variation on the landrace – not the landrace as a whole. It might be possible for future landrace variations to be handed over, by the individual farmers who created them, to their communities, as long as these communities were given corporate status in law – in the same way scientists hand over their rights to the plant strains they create to their employers. However, campaigners want whole landraces, the traditional varieties used by cultural communities for centuries, covered by IPRs.

In the previous section we referred to the WIPO/UNESCO Model Law on Folklore. It will be recalled that one of the significant aspects of this Model Law was that it recognized that cultural creations are not necessarily fixed and finalized but could be ongoing and evolutionary. Landraces are similarly ongoing and evolutionary; while the latest manifestation of a landrace may embody recent innovations by farmers, mostly it is a result of centuries of cumulative innovation and wise farming practice. So what is desirable from the community's point of view is that the landrace, as an ever-evolving but nonetheless identifiable variety, is protected, rather than just the latest manifestation of it. But the individual creators of a whole landrace are not identifiable, because they do not exist: there is no author; it is generations of the community who create landraces. However, this means that, under the entitlement theory, no identifiable entity exists that can be said to hold the moral entitlement to property rights in the landrace.

It is important to note that we have by no means established here that cultural communities *should not* be granted rights over landraces. We have merely shown that a labour theory of ownership is wholly inappropriate in accounting for this ownership. This is, at root, because the entitlement theory of intellectual property ownership needs a figure at the centre who can be identified unambiguously as the 'author' (or, in Drahos' terms, a 'first connection'), and there is no such figure in the case of landraces.

Entitlement Theory and Folk Knowledge

Claims are also made by some writers that cultural communities are entitled to IPRs in their traditional or folk knowledge; that is, their skill and expertise relating to the various properties of living resources and how to harness them for human benefit. As we explained above, indigenous knowledge has become much sought after by Northern firms and public institutions. Some writers have complained that indigenous knowledge is being plundered by western transnational corporations who are then patenting this knowledge for themselves. Vandana Shiva and Radha Holla-Bhar, for example, present a convincing account of how the American corporation W.R. Grace has hijacked Indian cultural knowledge of the properties of the neem tree and now holds patents based very strongly upon this knowledge (Shiva and Holla-Bhar, 1993, p. 224). Brush argues that traditional knowledge is useful and is therefore entitled to IP protection on the same basis as western scientific knowledge:

> If commoditizing Western scientific knowledge is justified by the wider public interest served, then indigenous knowledge

is likewise entitled to protection as intellectual property because it is useful in such areas as conserving biological diversity or identifying pharmacologically active plant compounds. (Brush, 1993, p. 659)

However, Brush is here illegitimately mixing utilitarian and entitlement theories of intellectual property. He is right to point out that one of the prime justifications for IPRs in the West, whether it is a valid justification or not, is that they increase the rate of innovation and thereby increase overall social utility. But the point is not that Western scientific knowledge is 'useful' and therefore 'entitled' to protection; rather that *legally protecting* western scientific knowledge is useful overall. The mere fact that cultural community knowledge is useful does not, on either an entitlement or utility theory, mean that it ought to be legally protected. An entitlement theory finds the consequent rise or fall in utility irrelevant, while for the utilitarian argument in favour of IPRs for communities to be valid, it would have to be shown not only that cultural community knowledge is useful, but that protecting cultural community knowledge through IPRs would increase overall utility.

There is another difficulty standing in the way of an entitlement justification of traditional knowledge. Under a labour-entitlement theory of intellectual property, it is a single act of creation[8] that generates an entitlement on the part of the creator to the exclusive benefit of the fruits of the creation. With traditional knowledge, however, there is no single act of creation: traditional knowledge is, generally, not the discovery of a single person or group of people, but is the result of centuries of collective experience – in which case there is never any one person or group of persons *entitled* to private property in this knowledge.

Moreover, even if there were an individual act of creation, certain elements of the entitlement theory entail that there can be no present entitlement to IPRs in traditional knowledge. Intellectual property is intended to give inventors and creators a monopoly on the fruits of their inventions and creations *for a limited time*. The time limit is not an arbitrary legal notion, but, if we adhere to the entitlement theory, is crucial, because it is recognized that invention is a partly social phenomenon: inventors rely on education, a political and social environment conducive to innovation, adequate materials, and so on, which are all provided socially (Spencer, 1970, pp. 140–2). It is also likely that others would have come up with the same invention sooner or later (Nozick, 1974, p. 182). For this reason, ideas and inventions turn into common property after a certain duration. In the light of these facts, it seems clear that traditional knowledge, such as the Indians have in the properties of neem, does not qualify as intellectual property on an entitlement theory. Even if traditional knowledge was, originally, the product of a single person or group, it seems that the time limit would have run out by now and the knowledge would have become common. It is certain, furthermore, that the individual creators/discoverers of such knowledge, if they existed at all, would now be dead. As IPRs are not inheritable,[9] an entitlement theory cannot ground property rights for the present members of cultural communities in the traditional knowledge of these communities.

In conclusion, we have not shown that we ought not to grant IPRs over genetic resources to cultural or local communities; merely that such communities do not hold *moral entitlements* to such property. Labour-entitlement theories are only applicable to communities if we subscribe to dubious organicist ideas of communities as independently existing entities ideas that, even if acceptable, preclude the possibility of communities ever being able to allow the outside use of the PGRs they supposedly own. Finally, it has been argued that the 'traditionality' of traditional knowledge – the fact that it is common knowledge, the product of collective experience without a single act of creation – precludes its being seen, from the point of view of an entitlement theory, as intellectual property. It seems, then, that entitlement theories cannot be meaningfully and consistently applied to communities. An entitlement theory of IP demands two things: an individual creator or group of creators, and an identifiable creative act. To use the terminology of the previous chapter, there has to be an identifiable author-figure (in the case of Boyle's theory) and a 'first connection', an 'act of personal demarcation' (as with Drahos). Neither of these requirements is satisfied in the case of communities and PGRs and their associated knowledge.

The fact that the entitlement theory does not justify community IPRs is, of course, no reason to decide that such rights are illegitimate. It merely shows that the entitlement theory is an inadequate basis for them. In the next section, we adopt an instrumental outlook on intellectual property and argue for community IPRs on the basis of autonomy.

SECTION III: AN AUTONOMY-BASED ARGUMENT FOR
COMMUNITY INTELLECTUAL PROPERTY RIGHTS

The previous section showed that an entitlement theory cannot justify granting intellectual property rights (IPRs) to cultural communities in either plant genetic resources (PGRs) or traditional knowledge. In this section, we argue for a theory that, we believe, compels us to adopt IPRs for cultural communities. The first part of the section presents this theory – an autonomy theory – for granting intellectual property rights to communities. This autonomy theory is based on the ideas of Will Kymlicka, the Canadian political theorist who has shown that special cultural rights can be derived from individualist, liberal, premises. The second part of the section suggests how the community intellectual property rights that are justified by the autonomy theory might take shape. It is argued that these rights include the right of exclusion and rent, but not rights of alienation.

Community IPRs from the Perspective of Individual Autonomy

The starting point for the argument advanced in this section is an idea aired by Tom Greaves in an article in which he tells us how

> tribal peoples have only their culture to distinguish themselves from everyone else. Their culture gives them their identity and their sense of value as a people. Disseminating that culture to outsiders dilutes their sense of personhood. (Greaves 1995, p. 203)

We fully endorse these sentiments and believe that they open up a new avenue of thinking about intellectual property rights for communities, taking us away from the flawed entitlement argument. The great virtue of linking cultural integrity with personhood is that it leaves behind quasi-mystical notions of communities as real entities, capable of labouring, and locates value in individuals. Cultural communities become valuable for their importance to individual lives and not in themselves. Greaves does not develop these ideas any further, however; as an applied anthropologist, his concern is more with working out practical ways in which indigenous communities can protect their cultural property than with political philosophy. More work needs to be done, then, for us to arrive at a satisfactory justification for community IPRs. In particular, we need to know: what is the value of 'personhood'? Why is membership in a cultural community necessary for our sense of personhood? What kind of legal rights are implied by such a theory? Do they include powers to protect cultural creations, including plant genetic resources and botanical knowledge, through intellectual property rights or similar legal provisions?

In order to find answers to these questions, we turn to the Canadian political theorist Will Kymlicka and his view of the relationship between individual rights and cultural community membership. Essentially, Kymlicka's argument, to be found in his book, *Liberalism, Community and Culture*, is that cultural communities are entitled to special protection under the law when their integrity is threatened by outside forces. This is not to say that communities are inherently valuable and are entitled to protection in themselves; rather, Kymlicka grounds his theory in individual autonomy. Respect for the equal right of individuals to the resources necessary to form their own conception of the good life entails that we guarantee their access to the cultural framework which is a precondition of forming such a conception. This means, under certain circumstances, that full respect for the individual requires the protection of their culture via special legal provisions.

It is our contention that IPRs in PGRs for communities are one of the special rights necessary under Kymlicka's proposals. We will explain why shortly. First, we give a more detailed account of Kymlicka's ideas.

Kymlicka: *Liberalism, Community and Culture*
Liberalism, Community and Culture is an attempt to reconcile what have become two opposing strands of thinking within political philosophy: liberalism and communitarianism.

'Communitarianism' is the name given to a set of diverse critiques of liberalism that surfaced in the 1970s and 80s in the wake of Rawls' *A Theory of Justice*. What all these critiques have in common is an aversion to what communitarian authors (principally Sandel, MacIntyre, Taylor and Walzer) perceive to be liberalism's theory of the self, i.e., as an 'antecedently individuated', atomized individual, seemingly formed from nowhere, unencumbered by constitutive ties to the community and prior to its ends as a person. In fact, say the communitarians, people's identities are formed by, and are inextricable from, both the cultural milieu in which they live and their own aims in life. The liberal idea of the self leads to treating people simply as citizens within a political community, ignoring their membership of real cultural communities. According to communitarians, this has a disintegrative effect on communities, which need some recognition in order to remain healthy and vibrant.

However, Kymlicka attempts to show that these two sides are not as far apart as is perceived, and that liberalism can accommodate many of the criticisms levelled at it by the communitarians. For our purposes his crucial achievement is to show how liberalism does not necessarily preclude the possibility of collective cultural rights, and in certain cases even requires such rights.

Kymlicka concurs with the communitarians insofar as he believes that liberalism's traditional emphasis on citizenship in a political community has led it to overlook the embeddedness of individuals in cultural communities. From this misunderstanding has followed a neglect of individuals' fundamentally 'cultural' character, which has often created a real-life tension: treating each individual only as an equal citizen in a political community can have a fracturing impact on minority cultural communities. We mentioned in the introduction to this chapter how the treatment of indigenous peoples was characterized until the 1970s by an attitude of 'assimilationism': integrating indigenous peoples into the wider society and downplaying their different background. In Kymlicka's own country, for example, a latter-day liberal crusade was launched at the end of the 1960s aimed at removing the special constitutional status of Indians and Inuit peoples. The liberal principle of equality was interpreted as implying a total disregard of people's ethnic

backgrounds; equality, it was thought, consisted in guaranteeing the same legal rights for each and every individual, whatever their cultural background. The trouble is that this 'loads
the dice' against members of minority cultures. For example,
guaranteeing the right of all Canadians, regardless of cultural
membership, to live, work and vote anywhere within the
national territory, might encourage the immigration of outsiders from the majority culture(s) into indigenous lands and
allow them to dominate such areas when they got there. The
constitutional changes provoked so much opposition from
indigenous groups – the very people whose equality they were
intended to respect – that they were withdrawn (Kymlicka,
1989, p. 156). These groups are now advocating legal rights to
protect their communities from the intrusion of outsiders
(Kymlicka, 1989, pp. 146–7).

Crucially, Kymlicka thinks that a putatively liberal 'colour-
blind' policy in fact goes against the liberal principle of equality: 'it ignores a potentially devastating problem faced by
aboriginal people, but not by English-Canadians – the loss of
cultural membership' (Kymlicka, 1989, p. 151). But why should
this be important to a liberal? The answer, according to
Kymlicka, is because one precondition for the development of
autonomy in the individual – and only an autonomous life is a
free one – is a rich cultural background. We will expand on this
idea by first turning to Kymlicka's opinion of the basis of liberal
autonomy.

Kymlicka argues that one of the most common criticisms
levelled at liberalism is that it is sceptical about matters
concerning the good life. He quotes Alison Jagger, who claims
liberals think that

> there are no rational criteria for identifying what is good for
> human individuals other than what those individuals say is
> good for them. Consequently, individuals' expressed desires
> are taken as identical with their 'real' needs, wants and
> interests. (Jaggar, quoted in Kymlicka, 1989, p. 17)

The version of liberalism presented and criticized by Jagger
sees all individual choices to be equally valid. This whole-
hearted scepticism is said to be the basis for the liberal emphasis on freedom of choice. The task of politics under this
liberalism is merely to ensure the conditions in which individu-

als can pursue their arbitrarily chosen desires. But, says Kymlicka, this is 'a complete misinterpretation' (Kymlicka, 1989, p. 18). In fact, the reason why freedom of choice is guaranteed within liberalism is that it is recognized that the only valuable life is one that is led from the inside. People have to make their own choices in life for that life to be one worth living, and 'If lives have to be led from the inside, then that freedom alone will justify the traditional liberal prohibitions on coercive paternalism' (Kymlicka, 1989, p. 18).

But the fact that people must make their own choices does not imply that those choices are equally valid. Kymlicka, referring to J.S. Mill (1972) – thus demonstrating that at least some liberals have never been sceptical over questions of the good life – says: 'Some projects *are* more valuable than others, and liberty is needed precisely to find out what is valuable in life – to question, re-examine, and revise our beliefs about value' (Kymlicka, 1989, p. 18). Liberty, then, is revealed not in libertarian terms as a simple absence of external constraint, but as genuine autonomy: the inner ability to make good choices about the good life on non-arbitrary grounds. This, incidentally, provides an answer to one of the questions we posed of Greaves earlier: what is the value of personhood? Autonomy is personhood; a fully formed person is an autonomous being and vice versa. The value of personhood, then, is that only as fully formed autonomous persons can human beings live a truly good life. But what is the relationship between personhood/autonomy and cultural membership?

According to Kymlicka, many things are necessary for the development and maintenance of genuine autonomy in the individual – education, freedom of speech and conscience, and most significantly for our purposes, a rich and secure cultural background. Only by living in a cultural community can people make informed, valuable choices about how to live their own lives:

> Liberals should be concerned with the fate of cultural structures, not because they have some moral status of their own, but because it's only through having a rich and secure cultural structure that people can become aware, in a vivid way, of the options available to them, and intelligently examine their value. (Kymlicka, 1989, p. 165)

All individuals have a right to the resources necessary to develop the inner ability to make good choices about which lifestyles are valuable to lead. A secure cultural background is one of these resources; each individual is entitled to grow up in, and live their lives in, a secure culture that gives their lives meaning. Liberal equality therefore implies that we ought to preserve cultures, and if this requires the institution of special cultural rights, like restrictions on immigration into tribal lands, this is what ought to be provided.

Such rights are only justified when a culture is in danger of disintegration from outside forces. This is because they are aimed at one thing only – the equality, in terms of personal autonomy, of cultural members with the rest of society. Members of the majority culture grow up and live their lives in a secure cultural environment: this culture is not under threat of extinction or dilution. In a liberal state dominated by 'colour-blind' policies that inadvertently work to damage minority cultural communities, the members of such communities are denied the security enjoyed by members of the wider culture. In Greaves' terms, their 'personhood' is diluted. This is a crucial inequality that cannot be ignored – indeed, must be rectified – by a liberal state. Special cultural rights are necessary in order that the members of minority cultures have the opportunity of developing a freely chosen life-plan commensurate with the shared understandings of their culture; an opportunity which is equal to the opportunity of the members of the majority culture. To institute special cultural rights for the majority culture is unnecessary and, indeed, could worsen the original inequality, and therefore has no justification.

An objection might be made to Kymlicka's argument on grounds that while he shows convincingly that everyone needs a rich cultural background in order to develop into an autonomous person with the capacity to originate and pursue their own conception of the good, he does not demonstrate why that cultural background must be that of their ancestors. Surely, it might be argued, a non-ancestral cultural background could be every bit as effective in providing a secure context in which individuals could develop their autonomy? On this view, what is needed is not necessarily a parental or traditional culture, but any supportive environment in which people could establish their own identity and give meaning to their lives.

However, Kymlicka is entitled to respond to this criticism by pointing out that we cannot simply invent backgrounds or contexts in which people can find their purposes in life. Such backgrounds are inevitably historically grounded. Where there is a community already in existence offering just such a secure background, we would be ill-advised to refuse to support it when it is threatened with disintegrative forces, since it may not be easily replaceable.

Community IPRs in Genetic Resource and Associated Knowledge as a Case of Special Cultural Rights

It is quite clear that cultural creations – such as works of art, sacred icons, types of clothing and living materials – often have a special significance for the individual members of the cultural communities out of which they arise. This does not imply that the communities are entitled to property rights purely by virtue of having created these things. The labour theory of property cannot coherently be applied to cultural communities, as we demonstrated in the second section of this chapter. Nevertheless, if cultural creations are lost, or are used in an inappropriate manner by non-members, the damage can be keenly felt:

> When a Hopi man or woman walks down a Tucson street and sees the mythic symbols, handed down from the elders, adorning a tourist's jogging shorts, culture dies a little – and with it, what makes that person a Hopi. (Greaves, 1995, p. 204)

But it is not simply a case of individual members being dismayed or offended by outside use of cultural creations; these creations can often be so central to the cultural community that we can talk of cultural integrity being compromised when control over them is lost and outsiders can use and abuse them with impunity. In such circumstances, the harm goes deeper than mere personal offence: when cultural integrity is weakened, so is the ability of future individual cultural members to form meaningful conceptions of their own good, and to live meaningful lives. For this reason, cultural communities must be invested with limited property rights in order to control the use of their cultural creations.

In our view, the cultural creations to be protected must include plant genetic resources and their associated knowledge.

To some, this might seem strange. Perhaps, it might be said, cultural integrity depends on the protection of religious imagery and works of folk art, but how can cultural integrity, and derivatively, individual autonomy, depend on the legal protection of a type of plant?

In answering this criticism, the first thing we would point out is that it is a highly selective view of culture that sees it merely in terms of outward expressions of the more obvious types of creativity, such as arts and crafts. Culture is quite clearly more than that. It is also about how people do the mundane things in life, like organizing their working day, socializing with their fellows, or growing food.

The second thing to be said is that the seemingly mundane practice of growing food is often the most 'cultural' activity of all, bound up with all kinds of beliefs about the world and the community's relationship to it. The Hopi Indians of southwestern USA, for example, believe their unique blue-coloured corn to be a gift from the 'Creator' to them specifically and therefore extremely sacred (Soleri et al., 1996). When outsiders use this corn in a crass, commercial way, the Hopi's deepest self-understandings are trivialized. We in the secular West may think of food simply as fuel, or sometimes as a type of sensual delight, but not all cultures share this view.

In any case, it is not necessary to show that the loss of control of traditional plant varieties would on its own result in the disintegration of communities like the Hopis. No one would suggest anything so simplistic. The point is that landraces and the knowledge associated with them are an important part of many indigenous communities' cultural structures. Removing any single part of this structure might not result in its complete collapse, but it would weaken the structure; for an outsider to weaken cultural structures cannot, in most cases, be justified.

The Extent of Cultural Communities' Rights over their PGRs

Having established that cultural communities ought to be granted certain rights over their PGRs, we can now begin to give an indication of how far these rights might extend, and what limits to these rights there might be. The rights to be granted to cultural communities in their PGRs are property rights in that they indicate legal powers over a thing. First, we will give an account of property rights to show why communities' rights over their PGRs are property rights; and second, we will indicate which property rights in PGRs the Kymlickian theory justifies.

Property rights

Some commentators have argued that community intellectual property rights are not a form of intellectual property, and indeed, are not a form of property at all. Shiva, for example, using the rubric of 'Community Intellectual Rights' to refer to the same sort of arrangements as community IPRs, writes that:

> The word 'property' has been deliberately excluded [by Shiva herself] in describing the knowledge systems of communities. Property rights in the term 'intellectual property rights' as presently understood, connotes commoditisation and ownership in private hands primarily for commercial exchange. (Shiva, 1996, p. 1630)

The reason why such commentators as Shiva deny that community intellectual property rights are property, is bound up with their general cultural relativism. They believe that the term 'property' has connotations of domination over people and nature that indigenous peoples would feel uneasy about. The Working Group on Traditional Resource Rights (WGTRR) says that 'property, for indigenous peoples and local communities frequently has intangible, spiritual manifestations, and, although worthy of protection, can belong to no human being' (WGTRR, 1997).

This worry over the word 'property' arises out of the anthropologists' tendency to relativize all normative concepts. However, words do not just have connotations; they have commonly understood meanings that can be validated. If we look at the meaning of property, it is clear that community intellectual property rights *are* a form of property. Shiva cannot pretend that her 'Community Intellectual Rights' are not a form of property simply by omitting the word 'property'. It is important to recognize that community IPRs are a form of property because it helps us to think more clearly about what kinds of rights indigenous and local farming communities ought to have over their traditional varieties and knowledge.

Ordinary property rights and intellectual property rights are essentially alike because they both indicate a relationship between persons concerning a thing. The fact that ordinary property rights refer to rights over tangible things and intellectual property rights refer to rights over abstract things is not significant. This essential similarity means that there is no reason to think that ordinary property rights and intellectual property rights have to be justified in different ways – say, entitlement for ordinary property rights and utilitarianism for intellectual property rights. The following discussion of property rights in general, then, applies as much to intellectual property rights as much as it does to property rights in corporeal things.

Since the work of Hohfeld (1919) it has been recognized that the lay understanding of property as a singular right is inaccurate, and in fact the terms 'property' and 'ownership' refer to a 'bundle' of rights that are logically distinct. Following Honoré (1961), Becker identifies 11 different rights (and liabilities) that make up ownership. These are the right to possess; to use; to manage; to receive an income; to consume or destroy; to modify; to alienate; to transmit; to security; to absence of term; and liability to execution (Becker, 1977, pp. 18–19). Honoré calls the state of affairs when every one of these rights is held with regard to something 'full or liberal ownership'. But ownership does not have to be full or liberal. Indeed, as Becker (1977, p. 22) points out, 'None of the characteristics which define the full or liberal notion of ownership in modern legal systems is necessary to all varieties of ownership.' In practice, full or liberal ownership is rare: there are many different systems of ownership in different times and places, nearly all of which fall short of full ownership. For example, owners may have the right to prevent others from building on their land, but not the right to prevent others from merely walking across it. To use an example closer to the subject at hand, most intellectual property rights do not have an absence of term, but elapse after a period of time. They are nevertheless property rights.

The fact that property is not itself a right, but a combination of many different rights, means that each of these rights must be justified independently. Property rights, including intellectual property rights, can be justified in one of two ways, directly or indirectly. The direct justification takes property rights to be primary, moral rights: a claimant is said to have a direct moral entitlement to something by virtue of a special relationship to that thing, such as having laboured on it. We have demonstrated elsewhere that, whatever the general validity of the direct entitlement theory, it cannot be applied in the case of communities and PGRs. The indirect, or instrumental, justification takes property rights to be secondary, legal rights, and can take one of two forms: utilitarian and moral. The utilitarian justification is that granting property rights will maximize the overall social good. This is probably the main popular justification for intellectual property rights in general, but it is useless in attempting to justify the granting of intellectual property rights to specific individuals or groups in specific things (i.e., what Becker calls the 'specific' justification of property rights). This is because, as Onora O'Neill points out, there is simply too much information to be processed, and too many calculations to be made, for us to be able to come to the solution that IPR actually does maximize utility (O'Neill, 1986, pp. 63–9).

The moral form of the indirect justification of property rights is that granting property rights in something to an individual or group will protect or promote some other right, such as autonomy. This is the form of justification we have adopted for community IPRs, using Kymlicka's ideas as a basis. We have not as yet, however, explained precisely what form of community IPRs are justified by our theory, i.e., which of the individual rights we associate with ownership can be justifiably granted to cultural communities in their plant genetic resources and associated knowledge. This is the question to which we now turn.

Extent of, and Limits on, Cultural Communities' IPRs
Which of the rights that go to make up the bundle we call property can justifiably be granted to cultural communities in their PGRs, and which denied? As we have seen, traditional varieties and botanical knowledge are to be protected through intellectual property rights because they are a crucial part of the cultural structure in which individuals grow up and live their lives. This indicates the distinction we need – namely that the legal rights to be granted over communities' varieties and knowledge must be only those which promote cultural integrity and prevent, as Greaves puts it, the 'dilution' of the culture

(Greaves, 1995, p. 203) and subsequent damage to the autonomy of its individual members. We believe that these rights
must include rights of possession (exclusion), use, management, the right to receive an income (to rent, i.e. to sell rights
of usage to non-community members), and (unusually for intellectual property rights) absence of term. But they do not
include, to use Honoré's terms, the right to alienation (sale),
and the right to destroy. We will now explain why the autonomy-based theory allows or prohibits these rights, leaving the
discussion of the right to rent until last, because it is the most
interesting from our point of view.

The rights of possession, use and management are obviously
justified by the theory. Since the central claim is that communities ought to be able to control the outside use of the things
that are important to their integrity; the right of possession is
the most important right. Rights of use and management speak
for themselves. Intellectual property rights usually include a
limit on their duration; patents in the UK run for 17 years, for
example, while copyrights last as long as the author lives, plus
another fifty years. There are good reasons for these stipula-
tions in each case, but a time limit is not a necessary corollary
of intellectual property rights *per se*. Clearly, in the case of community IPRs, such a limit would defeat the purpose of granting
intellectual property rights to cultural communities, which is
to maintain communities as going concerns in perpetuity.
Community IPRs must therefore enjoy the absence of term
shared by most ordinary property rights.

The prohibition on destruction is surely self-evident; the
reason to grant rights in the first place is to protect a community's integrity. To allow the destruction of the community's
PGRs by one generation would compromise the position
of future generations. The other prohibition, on alienation, is
more controversial and therefore must be explained in greater
detail.

The whole reason to grant cultural communities rights over
their PGRs in the first place is so that they can maintain
control over the things that matter to them as a community,
thereby ensuring a meaningful continuing existence and a
secure structure in which individuals can develop. To grant the
power to present generations of a community the power to sell
the rights to their cultural creations on the open market would

undermine that objective: if the cultural creations were alienated from the community, future generations of the community
would be growing up in an impoverished culture. This is not to
deny indigenous communities all the other rights that other
members of society have; we must remember that only cultural
communities that are endangered are entitled to special legal
protection in the first place. Members of indigenous communities would have all the other individual rights that others in the
wider society have, including the right to buy and sell property
that they own individually and in conjunction with others –
though not essential resources such as land. As Kymlicka points
out, dividing up communal goods among individual members
and 'granting' them sale rights, thereby subjecting them to economic pressures, was a tactic often used by the Europeans to
'legitimately' gain title to indigenous lands (Kymlicka, 1989,
pp. 147–8). To deny cultural communities the power to turn
their cultural creations into goods to be bought and sold is only
to deny them the power to destroy themselves.[10]

Any rights granted to PGRs must, then, be non-alienable.
This does not necessarily mean that the right to *use* PGRs
cannot be given to outsiders, either free or in return for
payment. Such an arrangement would not result in the permanent loss of control over a creation that selling off the rights
altogether would. However, we do have certain misgivings about
granting the right to 'rent' PGRs, i.e., allowing the practice of
selling use rights. We will now explain the nature of these
doubts.

The extra-cultural use of cultural creations like plant genetic
resources and knowledge has the potential to damage a
culture's sense of itself, in turn damaging the context of choice
in which individual members live their lives. If there was no
such danger, there would be no justification for granting any
rights over cultural creations. Our worry is that if cultural communities were granted the power to allow the use of cultural
creations in return for money, even without permanently alien-
ating their control, they might sometimes allow the use of their
creations when it is against their deepest wishes. In short, they
might be susceptible to offers they could not refuse. We earlier
mentioned the colonial practice of granting American Indian
communities property rights over their land and then buying
the land in order to break up communities. Such communities,

despite wanting to remain in existence, were often impoverished by colonialism and found the prospect of the money they were being offered impossible to turn down. It seems to us that allowing communities the right to 'sell' rights to use their resources and knowledge might bring similar dangers: one can easily imagine cases where a cultural community would prefer not to allow any outside access to certain aspects of its heritage, but if the alternative is impoverishment, they might feel they have no choice. The result would be the degradation of the cultural context. Given that the protection of this context is the sole justification for special rights in the first place, can we justify granting rights that might have the effect of degrading it?

However, there is another perspective on this issue. It might be objected that prohibiting the sale of use or access rights would condemn cultural communities to permanent impoverishment, which would in turn lead to the break-up of that community. It might be possible for a cultural community to remain intact despite poverty, but if this community finds itself in a wider community where the opportunities for material benefit are greater, the chances are that the community will disintegrate, losing its special character as its members join, perhaps reluctantly, the majority culture. We would suggest, however, that this alone does not justify granting cultural communities the power to sell rights of use over their creations. Other things being equal, the obligation of the wider political community to ensure that cultural communities remain viable, means ensuring that pieces of cultural heritage do not have to be 'sold off' in order for cultural members to escape poverty. Financial help should be given and perhaps certain sorts of affirmative action programmes should be instituted to ensure that this is not necessary.

Unfortunately, many minority cultural communities around the world are to be found in poor countries. These countries cannot afford to preserve their minority cultural communities exactly as they would wish. In such situations, an agreement like the recent Kani/*jeevani* contract,[11] allowing minority cultural communities to receive financial benefits in return for sharing their PGRs or their botanical knowledge might be the best available option. They might feel that a little part of their culture has been lost to them, but at least that will have helped to ensure their continued existence, albeit in slightly degraded form. Not allowing them to benefit financially might condemn them to disintegration. For this reason, rights of rent are, we believe, justified by the autonomy-based theory we have set out.

Cultural communities are entitled to the right to exclude others from the use of their resources and knowledge should they so wish. However, this right is not absolute. In ordinary circumstances, a community ought to be able to refuse completely the outside use of its resources and knowledge. In most cases, the community has the greatest interest in their resources – they form part of the cultural heritage, part of the context in which individual lives are led. Outsiders' wishes will usually not be significant enough to override the desire of cultural community members to keep their heritage to themselves. Nevertheless, there are certain circumstances in which we can imagine the community's wishes being justifiably overridden. To use the most obvious example, many medicines are developed from plants known to and grown by cultural communities. Millions of lives are prolonged every year through the use of such medicines. There are probably many more medicinal plants used by cultural communities that western medicine does not yet know about; indeed, prospecting for medicinal plants is a growth industry at the moment. It is quite conceivable that in the near future a plant will be discovered by western scientists which they are eager to remove from the community in which it is found in order to develop into a modern drug. Would the community be entitled to prevent the removal of this germplasm, or its commercialization, on the grounds that this would weaken its integrity? We believe most people would intuitively think not, and we also believe this intuition is justified. As we said above, in ordinary circumstances, the community's interest in a particular resource outweighs the interest of outsiders. But in the hypothetical case outlined above, the interest of a large number of outsiders in continuing to live is clearly more significant than the interest of cultural community members in what will probably be only a relatively small contribution to their cultural integrity.

This case demonstrates that the argument from autonomy can only take us so far: it can ground community intellectual property rights, but at a certain point it comes up against another principle – the right to life of outsiders – which 'trumps' it. In cases like these, something like a compulsory purchase order could be decreed; the community could still be compensated,[12] but it would lose its entitlement to prevent outside use of the resources. The right of control, then, is not absolute. Again, this is not to deny property rights to cultural communities that are enjoyed by other individuals and groups. Compulsory purchase orders exist in most countries around the world, and are a recognition that property rights are never absolute: the whole community, and not just individual property owners, have an interest in any private property and this interest can sometimes override the interest of the owner. These principles imply that some agency has to be instituted to make the decisions over whether communities are entitled to control of plant genetic resources and knowledge in specific cases. We assume that this agency would need to be an international one, set up by international agreement to adjudicate such matters.

It might be argued that to prohibit alienation, sale of use rights, and destruction of PGRs by the communities that own them is a mockery of the idea of property; that any notion of property must involve at least the right to sell, and to rent. But this is not so. As Becker points out, alienability is not central to the notion of property at all: 'free alienability of property, and the power to dispose of one's holdings through a will, are rather recent accretions to the notion of ownership – neither natural nor necessary to human society per se' (Becker, 1980, p. 207). Many property systems, in the past and in other cultures in the present, do not include in the notion of property the right to alienate for material gain. There is no single right which constitutes the 'core' of the bundle of property rights; Becker claims that when any one of the rights we associate with property is combined with security, we can speak of a property right (Becker, 1977, p. 20). So we can still refer to plant genetic resources and botanical knowledge, indeed all cultural properties, as 'property' even if the communities designated as owners do not have rights of sale or of rent. They are acknowledged as owners because it is recognized that their interest in the thing in question ordinarily outweighs the interest of outsiders.

In conclusion, the argument from liberal autonomy, inspired by Kymlicka, demands that we grant intellectual property rights to cultural communities in their plant genetic resources and botanical knowledge. By rooting community rights in individual autonomy, we avoid the incoherence of theories that seek to portray communities as singular entities, capable of holding entitlements to property *qua* groups. Rather, community IPRs are recognized as secondary, legal rights which are justified by the contribution they can make to the ongoing integrity of communities that might otherwise be endangered. In turn, the reason that it is important that communities remain in existence is that community members derive from their membership the resources necessary to become and remain freely choosing autonomous agents.

But, as we have shown, communities' intellectual property rights must be limited in two ways: first, they do not include some of the rights we normally associate with property, principally the right to alienation. This is because these rights are not justified by the autonomy principle; they would make no contribution to, indeed would detract from, communal integrity, and thus would endanger the autonomy of cultural members. Secondly, the rights to be granted can justifiably be overridden in certain circumstances, namely, when the interest of humankind in using a community's resources or botanical knowledge outweighs the interest of that community in keeping them exclusive. In these circumstances, compensation would be payable to the community in question.

CONCLUSION

We have argued that intellectual property rights for communities are desirable from the point of view of preserving indigenous and farming communities. But there still remain some questions to be answered regarding community IPRs. The first is whether community IPRs are really necessary. We already have a well-worked-out system of intellectual property rights that, thanks to the 1994 GATT agreement, is now an entrenched part of the international trading environment. It might be that indigenous rights to intangible objects could be incorporated into one of the standards forms of intellectual property rights – patents, copyrights, trademarks, trade secrets and plant breeders' rights – without the need for a brand new *sui generis* system (da Costa e Silva, 1995). If this could be done, then it would avoid all the difficulties normally associated with the creation of a new intellectual property form. The painstaking progress of the process for creating a new international system for semiconductor chips is a classic example of these difficulties (Cornish, 1993, p. 55).

However, it seems doubtful that any of the present forms of intellectual property could accommodate intellectual property rights for communities. As Kerry ten Kate writes, '[c]urrent legal regimes are not well suited to protect indigenous innovation' (ten Kate, 1995, p. 19). The most obvious candidates for protection of traditional plant varieties would be patents and plant breeders' rights. But the requirements for granting both patents and plant breeders' rights are too stringent for traditional varieties to be 'let in' to the system. Let us look at these systems in turn.

For an invention (now including plants) to be patented it has to be novel, involve an inventive step, and be capable of industrial application (Phillips and Firth, 1994, p. 44; Bainbridge, 1995, p. 247). Moreover, a patent application must include a description of the invention for a court to understand, and so that anyone 'skilled in the art' in question could make it from the instructions provided. Traditional varieties can certainly claim to be novel, and to involve inventive steps, since they are constantly changing and evolving, and farmers who grow them continually experiment and innovate with them (RAFI, 1996b, p. 45). Traditional varieties can also claim to be capable of industrial application; indeed they are crucial to Northern agribusiness in that they form a bedrock of genetic variability which can be drawn upon to bolster Northern elite crop varieties. However, it seems unlikely that the process of creating a new variation on a traditional variety could be described at a level technical enough, first to satisfy a patent court, and second for a person skilled in the art to produce the variation from the same original materials.

Even if traditional varieties did turn out to be patentable, however, it is undesirable that such varieties should be protected through the patent system. Patents are limited in time; this is a sensible and indispensable aspect of patenting. It signifies that a bargain has been struck between society and inventor whereby a monopoly is granted to the inventor on the economic exploitation of his or her invention in exchange for the inventor making the invention publicly available and publishing details of how to make it. But a time limit on a community intellectual property right would make no sense. Its *raison d'être* is not to act as a reward for/incentive to innovation, or to respect entitlements created through labour, but to protect cultural communities. Any protection must therefore be permanent.

As for plant breeders' rights, under the UPOV system, for a plant variety to be protected it must be new, distinct, uniform and stable (da Costa e Silva, 1995, p. 548; Phillips and Firth, 1994, pp. 356–7). Even if traditional varieties could be shown to be new, distinctive, and stable, they are certainly not uniform – a relatively high degree of genetic variability is typically exhibited by any particular landrace, in terms of height, yield, disease resistance and so on. In fact, the value to humanity of traditional varieties lies precisely in their genetic variability. Northern 'agribusiness' is based on a very narrow genetic base; the narrower the base, the more the system is vulnerable to the vicissitudes of nature. Agribusiness needs the variability of traditional varieties; without them it would collapse. Nevertheless, this variability means that traditional varieties could not be incorporated under an intellectual property system that is designed to exclude variability.

What of the remaining forms of IPRs – copyrights, trade marks and trade secrets? Neither copyrights nor trade marks are appropriate, since copyrights lapse in time, while trade marks merely protect the brand name, not the product itself, from appropriation. This leaves trade secret legislation. However, despite the fact that trade secrets are usually described as a form of intellectual property, the legislation surrounding it is unlike the other forms in that it is not connected to particular abstract objects but is more concerned with protection against industrial espionage. There seems no reason why, as long as communities acquired legal personality, their knowledge could not be protected by trade secret legislation, but this is hardly the point. It is the use by outsiders that provides the *raison d'être* for community IPRs in the first place. Indigenous communities do not generally want to keep their knowledge secret (although if they do, they are, under the autonomy-based theory set out above, fully entitled to do so), but to be recognized as the holders of knowledge and to be compensated when it is used by outsiders to develop commercial products.

The main forms of intellectual property that are already recognized in law are, then, unsuitable for protecting traditional varieties. However, a second argument against the necessity of community intellectual property rights is that they are unnecessary at a time when bilateral contracts are providing indigenous communities with a measure of implicit intellectual property rights over plant genetic material and botanical knowledge.

A number of agreements have already been made between transnational corporations and other Northern groups interested in indigenous knowledge, and indigenous groups. Examples include Shaman Pharmaceuticals, a company set up to identify promising plants from the knowledge of traditional healers, and which has its own 'Healing Forest Conservancy' to facilitate a flow of benefits to the indigenous peoples concerned and to support conservation, and Sabinsa Corporation, a profit-oriented company hoping to introduce the traditional knowledge of India to North America. The Kani/*jeevani* agreement mentioned earlier in this chapter is another example of a bilateral contractual agreement that does not explicitly confer intellectual property rights on a community.

It might be argued that agreements such as these demonstrate the pointlessness of a specific *sui generis* intellectual property regime for traditional varieties and knowledge. Indigenous communities are already getting good deals from companies and other bodies interested in their knowledge, so what additional benefit would a *sui generis* regime provide?

There are five good arguments against relying purely on bilateral contracts. The first is that they do not provide any protection against the possibility of an outside interest getting hold of a traditional variety without the permission of the community from which it comes. If a company was able to access such a variety, the community would have no power to control the use of that part of its cultural heritage. The second argument is that bilateralism tends to work in favour of the most powerful interests. There are a relatively small number of multinational companies involved in bioprospecting, and a relatively large number of indigenous and farming communities. Bilateralism on its own might have the effect of 'playing off' communities against one another, with one community winning a bioprospecting contract on the basis that it offered a lower price.

The third argument is connected to the last point, and it is that often, two or more different communities share the same botanical knowledge and perhaps even traditional varieties. Bilateralism would result in all the benefits accruing to the community that was fortunate enough to be contacted by the outside interest. A community IPR regime would be able to designate ownership of the same knowledge or variety to two or more different communities, thereby ensuring that all the groups that share the same knowledge can share in the benefits. The fourth argument against bilateralism is that in the case of a dispute, the community would have little chance against a mighty transnational company with an army of contract lawyers behind it. A scheme of intellectual property rights for communities, on the other hand, if it were internationally administered, could offer assistance to indigenous communities that found themselves in conflict with bioprospectors.

The fifth argument for community IPRs as opposed to bilateral contracts refers back to our reasons for instituting community IPRs in the first place. Community IPRs are special rights that are aimed at preserving the integrity of cultural communities by giving them control over their cultural creations. One of the ways intellectual property rights, along with other sorts of special rights, do this is by recognizing that particular resources are the creation of particular communities. Through law, the wider community identifies a particular object – abstract or tangible – with a community. This general act of recognition by the international community cannot be provided by a bilateral contract.

Intellectual property rights for communities are, then, necessary as well as desirable and justifiable. They form one of the three challenges the developing world has made to the growing intellectual property proprietarianism of the industrialized world. The compatibility of intellectual property rights for communities with the other two of these challenges – the national sovereignty principle and the common heritage ethic – are dealt with in the next chapter. We argue that as long as these three principles are conceptualized correctly, then they are compatible.

However, despite agreements such as the Kani/*jeevani* contract and the activities of companies like Shaman Pharmaceuticals and the Body Shop, community intellectual property rights remain an aspiration rather than a reality. This is perhaps unsurprising. The developing world has been keen to press home the national sovereignty and common heritage principles – particularly the former – because it is obvious that Third World states benefit as states from the recognition of these principles. However, many developing countries' relations with their indigenous peoples are notoriously strained. States' innate fear of internal fragmentation has meant they have historically been reluctant to accord any special rights to groups that define themselves apart from the mainstream political community. Nevertheless, many developing countries now recognize the value of according such rights to their indigenous groups. Community IPRs are on the agenda at various international fora, and it may not be long before some kind of regulatory system is established to recognize them.

NOTES

1. Following Fowler and Mooney, we ought to issue a justification for our use of the phrase 'traditional varieties'. To a modern plant breeder, a variety is something quite distinct, bred for a high degree of genetic uniformity. Traditional varieties display, on the other hand, a high degree of genetic diversity. There seems no other adequate phrase (in the past, the phrase 'primitive cultivar' has been used, and indeed still is in some quarters, but it seems rather insulting to Third World farmers so we have avoided it). We also use the word 'landrace' from time to time. As long as the distinction between modern breeders' varieties and traditional varieties is understood, we believe the term is admissible (Fowler and Mooney, 1990, p. xv).

2. There are, in fact, several suggestions for compensating indigenous communities for the use of their resources and knowledge. Some involve attributing intellectual property rights and some do not. There are also several ways in which *sui generis* intellectual property rights systems for communities have been conceptualized (ten Kate, 1995, p. 21). We take the view that specific rights ought to be granted to communities, and that these rights can legitimately be called intellectual property rights. Our reasons are elaborated in the third section of this chapter.

3. The campaign for community IPRs cannot be characterized as merely an adjunct of the indigenous rights movement, however. As set out in this book, it is a response to the perceived exploitation of indigenous and non-indigenous farming communities with regard to their

germplasm and botanical knowledge. It therefore resides as much in the discourse on plant genetic resource control as in the indigenous rights discourse.

4. The difference is that 'indigenous' peoples are the descendants of peoples who originally inhabited lands that are now controlled by the descendants of European settlers (and descendants of their slaves) such as the USA, while 'non-indigenous peoples', while often being 'indigenous' in the literal sense, inhabit states that were never given over so fully to settlers. Indian farming communities are therefore examples of non-indigenous communities. Greaves believes community IPRs should be restricted to indigenous communities and should not be extended to non-indigenous farming communities. His reasons are pragmatic – he believes dominant societies feel they have little to lose by granting special rights to their indigenous communities, but that 'When rural agricultural peoples are included, anxiety will rise that the dominant society's way of life may be disrupted' (Greaves, 1996, p. 12). However, this seems intuitively unfair, and, as we shall see in the next section, the autonomy-based justification for community intellectual property rights that we develop suggests that community IPRs should apply to non-indigenous communities as well as to indigenous ones.

5. The dividing line between Farmers' Rights (FRs) and community IPRs is confused. Originally the concepts were quite distinct, but some see community IPRs as part of FRs. The 1992 Convention on Biological Diversity, for example, seems to include Community IPRs in its conception of Farmers Rights when it recommends the extension of Farmers' Rights to indigenous communities (Article 9). However, we believe the concepts ought to be kept distinct for the sake of clarity. FRs refers to a non-specific compensation mechanism aimed at conservation, while community IPRs refer to a specific *sui generis* intellectual property system aimed at compensating particular indigenous and farming communities for using the traditional varieties and knowledge that have been specifically acknowledged as arising from their culture.

6. Shiva is here glossing over the difference between the principles of national state ownership and local community ownership of genetic resources. There is a prima facie distinction between ownership by states and ownership by local communities; the question of whether nation states do, or should, own these resources is a problem that is addressed in the next chapter on national sovereignty.

7. Strictly speaking, according to Locke, people are 'owned' by God, but a secular version of the Lockean theory entails self-ownership.

8. The phrase 'single act' should not be taken too literally. We use the phrase not to indicate a moment of inspiration – the 'Eureka!' moment, perhaps – but just that a single invention has been made, or that a single piece of literature has been written.

9. At present the only kind of IPRs that could conceivably be said to be inheritable are copyrights, which tend to run for the author's life plus a fixed number of years, depending on the country. The posthumous royalties are due to the author's estate, which usually means the family

(Hughes, 1988, p. 323). Patents and other IPRs are most certainly not inheritable; to make them so would be wholly novel.

10.   The objection is only to the selling of cultural creations and resources, including IPRs in PGRs, and not to other schemes such as the opening up of cultural sites to tourists, like that of the Acoma pueblo of the US (Greaves, 1995, p. 210), or to the commercial production of a community's landraces. The difference is that allowing outsiders to view sites, or producing one's own plants commercially, does not entail the permanent loss of a creation or resource important to the integrity of the cultural community. Selling the rights to a type of corn would entail this loss. Some communities may feel that even allowing outsiders to see their sacred sites will be harmful to them, or may be the 'thin end of the wedge'. If so, they would, under this proposal, reserve the right not to allow tourism. The crucial point is that control must reside with the community concerned.

11.   The Kani tribe of Kerala state, India, have recently been granted intellectual property rights in the plant *jeevani*, thought to have ginseng-like qualities. They led researchers voluntarily to the plant and will now receive a one-off fee and a 2 per cent share of any profits made (*Nature*, 16 May 1996, p. 182).

12.   It might be said that compensation misses the point: that under the theory presented here, what is important is not financial benefit but the ability of communities to protect their agricultural and botanical heritage from the outside forces that threaten the communities' integrity. But in situations where the retention of control by communities of their heritage cannot be justified, compensation at least recognizes that the community has lost some of its uniqueness. Financial compensation cannot directly replace this loss, but is at least a recognition of it. This is a principle we acknowledge generally; when people are killed in disasters, for example, we routinely compensate their relatives. This does not imply that the money is any replacement; it is merely a recognition of loss.

5 National Sovereignty

INTRODUCTION

We have argued that there are four basic principles that run through, and have a significant influence on, debates and practice concerning access to and ownership of genetic resources. In the previous chapters we have examined two of these principles – proprietarian intellectual property rights and community intellectual property rights. In this chapter we look at the influence of the third principle, that of national sovereignty. The chapter is divided into two sections. In Section I, we explain how and why sovereignty over natural resources is an entrenched principle in international law, and ask why, given this fact, the principle was not applied to genetic resources until relatively recently. In Section II, we examine the relationship between the principle of national sovereignty and the principle of community IPRs.

SECTION I: NATIONAL SOVEREIGNTY OVER NATURAL AND GENETIC RESOURCES

National sovereignty over natural resources forms part of the canon of customary international law (Correa, 1995). It is an unquestioned cornerstone of relations between states.[1] Indeed, it could be said that sovereignty is meaningless without control over natural resources. There are three principal international documents affirming sovereignty over natural resources. The first, and most important, of these is the General Assembly Resolution on Permanent Sovereignty over Natural Resources, passed on 14 December 1962. It affirms, unambiguously, the 'right of peoples and nations to permanent sovereignty over their natural wealth and resources', which it considers to be 'a basic constituent of the right to self-determination.' (UN GA Res 1803 (XVII)). Similarly, both the International Covenant on Economic, Social and Cultural Rights (993 UNTS 3) and the

International Covenant on Civil and Political Rights (999 UNTS 171) affirm the right of peoples to freely dispose of their natural resources. The later Charter of the Economic Rights and Duties of States says that sovereignty over natural resources is the right of 'every State' (Art 2 (c), GA Res 3281 (XXIX)).

There is a level of ambiguity, however, over the question of precisely where sovereignty over natural resources lies. Later in this chapter we discuss the relationship of the national sovereignty principle to the principle of intellectual property rights for indigenous and farming communities. Some writers believe the use of the word 'peoples' in such documents as the UN Charter and the Resolution on Permanent Sovereignty over Natural Resources indicates that in international law, indigenous peoples can be regarded as self-determining and therefore as possessing sovereign rights over the natural resources that are found within their territories. However, the dominant interpretation of international law – the one generally adhered to by governments – accords self-determination only to whole populations within territories where there is settled and stable government.

But what *moral* justification is there for national sovereignty over natural resources? Two justifications spring to mind. The first justification is founded on the claim that every nation has a right of self-determination – that is the defining basis of the nation state – and that to exercise that right of self-determination, the nation must have control over all the natural resources within its territory. This is regarded by some writers as a self-evident truth. As Ian Gambles (1990, pp. 40–1) puts it, 'The natural resources of the political community are constitutive of the community, which does not need to justify its possession of them ... It is the separation of a political community from its territory which requires justification, not the opposite.'

However, there are two difficulties with this argument; for one thing, there is a notorious lack of coincidence between 'nation' and 'state'; many nations are not states (e.g. Kurds) and many states are made up of more than one nation (e.g. UK). For another thing, it is not self-evident that the state 'owns' the territory it occupies, still less that it 'owns' all the natural resources within that territory. The location of territory and of natural resources is a matter of geographical accident, and carries no moral significance in itself. For example, the fact that some North Sea oil fields are located within the UK's 200-mile exclusive economic zone does not necessarily indicate any ethical claim by the UK to ownership of those oil fields. Morally speaking, it could be argued that such oil deposits belong to humanity as a whole or to all North Sea states.

The second justification for natural sovereignty over natural resources is instrumental; that in order to prevent disputes and conflict breaking out over the right to use those resources, every state must be given the exclusive right to control access to its natural resources. However, this justification is also flawed: rivalry between nation states has often resulted in wars over natural resources. Moreover, the justification is, at best, only consequentialist in character: even if national sovereignty over natural resources has served to maintain international peace, that would be only a prudential, not a moral, reason for accepting the principle of natural sovereignty.

Furthermore, even if the principle of national sovereignty over natural resources is deemed to be morally acceptable, it necessarily carries with it certain restrictions. That is to say, a nation state is not entitled to make use of its natural resources in such a way as to inflict injury on the inhabitants of another nation state, since that would be to violate the principle of national sovereignty itself. This restriction is commonly recognised by international law and conventions. For example, the 1992 Rio Declaration on Environment and Development proclaimed that:

> States have, in accordance with the Charter of the United Nations and the principles of international law, the sovereign right to exploit their own resources pursuant to their own environmental and developmental policies, and the responsibility to ensure that activities within their jurisdiction or control do not cause damage to the environment of other States or of areas beyond the limits of national jurisdiction. (Grubb et al., 1993, p. 87)

Clearly, then, the principle of national sovereignty over natural resources cannot be categorically asserted. Nevertheless, it has assumed the significance of a long-entrenched principle of international law, which makes it all the more surprising that national sovereignty over *genetic* resources was not generally accepted until very recently. This is due to the unusual nature of genetic resources compared to other kinds of natural resources. In economistic terms, genetic resources have public good qualities which means that the same unit can supply many additional customers at a negligible cost (Boyle, 1996, p. 12). This is because genetic resources contain information, in the form of DNA, that makes them able to reproduce themselves. It is this information that makes genetic resources valuable. So, at least in theory, only a small sample of genetic resources need be taken in order, eventually, to generate large economic returns (Subramanian, 1992, p. 105). As David Wood points out, the proportion of germplasm taken compared to the bulk remaining is minuscule (Wood, 1988, 277). For example, the entire coffee industry of South America is derived from a single tree taken from Sri Lanka in the early eighteenth century (Fowler and Mooney, 1990, p. 180).

Most other (non-renewable) resources are different in that they do not contain information and cannot therefore reproduce or grow. The amount of coal extracted from the earth, for example, is the same as the amount sold. We cannot copy coal, so appropriating small quantities of it is pointless, whereas a small quantity of genetic resources is all that need be appropriated in order to generate larger quantities and subsequent profits (Subramanian, 1992, p. 106). National sovereignty over coal is therefore easy to conceptualize, but it seems that the reproducibility of genetic resources for a long time blinded both germplasm collectors and biodiversity-rich countries to the possibility that these resources might also be considered subject to national sovereignty. Countries genuinely believed that in allowing collectors to remove samples, they were losing only trifling amounts of resources – a few seeds, in most cases. In fact they were losing control of a valuable information resource.

For three centuries before the 1970s, germplasm was collected with impunity by Northern prospectors, and this germplasm contributed an enormous amount to the agriculture of the developed world. Kloppenburg and Kleinman claim that it is 'no exaggeration to say that the plant genetic resources received as free goods from the third world have been worth untold *billions* of dollars to the advanced industrial nations' (Kloppenburg and Kleinman, 1987, p. 19). Under colonial conditions, there was nothing the colonies could have done to prevent the transfer of genetic resources to the European powers (and later the USA). However, even after the colonies achieved independence, it seems that it did not occur to countries possessing high levels of biodiversity how much they were losing by allowing prospectors to remove samples of their genetic resources (Walden, 1995, p. 180).

As we explained in Chapter 2, the ethic of free access and distribution prevailed until the North started protecting their own botanical innovations through intellectual property rights. At this point, many countries in the South began to grasp the situation; the propertization of the information contained in genetic resources alerted developing countries to the fact that information was what they were losing. Some countries had, in fact, operated embargoes on certain genetic resources for many years – the prime example being Ethiopia's coffee ban. Ethiopia is a centre of genetic variability for coffee, but for many years it refused to allow Northern researchers access to this variability because it did not want Northern coffee plant breeders to grow rich at its expense (Fowler and Mooney, 1990, p. 104). Ethiopia exercised *de facto* sovereignty over the coffee resources within its borders, but even in that situation, if any coffee plants had been taken from Ethiopia without permission, the lack of any accepted norm within the international community of national sovereignty over genetic resources meant that Ethiopia would have had no redress.

The 'seed wars' of the 1980s changed this situation, and now national sovereignty is an accepted principle in any international dealings over genetic resources. As we noted in Chapter 2, the 1983 version of the International Undertaking on Plant Genetic Resources declared genetic resources to be the common heritage of mankind; this was taken to mean that all genetic resources should be available without restriction. National sovereignty clearly played no part in this conception of common heritage (although, as we argue in the next chapter, national sovereignty is, in fact, compatible with a fuller interpretation of common heritage). Nevertheless, as the developing countries came to realize that common heritage – conceived as free access to all kinds of genetic resources – was a pipe-dream, they also realized that national sovereignty formed a part of the best solution they were likely to get. Accordingly, the 1991 revision of the International Undertaking on Plant Genetic

Resources declared that common heritage was subject to the sovereignty of states over their plant genetic resources (Correa, 1995, p. 60). A year later, Article 15 of the Convention on Biological Diversity acknowledged that 'the authority to determine access to genetic resources rests with the national governments'. As Grubb et al. (1993, p. 79) commented on the CBD, 'The negotiations therefore established that genetic resources can no longer be regarded as a common resource.'

It is important to point out that the application of the principle of national sovereignty to genetic information is consistent; indeed, it was inconsistent not to apply it to such resources. Information may be an intangible resource, but it is no less of a natural resource for that. The non-application of the principle of national sovereignty to genetic information was a convenient method by which the North was able to extract value from the South free of charge. It was clearly unfair and inconsistent. It is also important to show that the national sovereignty principle is compatible with both community intellectual property rights and the common heritage ethic. Indeed, we will argue in the next chapter that, although there are potential areas of conflict between community intellectual property rights and the national sovereignty principle, these difficulties are not conceptual and can be overcome; while in the case of common heritage, the common heritage ethic *requires* national sovereignty in order to be fulfilled. What we must now do is to show how national sovereignty can benefit developing countries.

National sovereignty over genetic resources is crucial for developing countries because it means they can control the outflow of these resources from their territory. Previously, countries could use their *de facto* power over their own territory to try to ensure that no genetic resources were smuggled out (although most countries did not, as they perceived it to be in their interests to adhere to the norm of free access; this norm has now, of course, been rejected). But the lack of an accepted norm of national sovereignty over genetic resources meant that there was nothing countries could do *de jure* to stop non-nationals from using the genetic resources as they wished. Economically, national sovereignty is highly beneficial because it means that countries can benefit from their biological diversity in a way they could not before. They can use the acceptance of the national sovereignty norm to negotiate access deals and material transfer agreements in order to receive income and other benefits such as technology transfer and training. Biodiversity is the primary asset of many developing countries and it is only fair that they should be able to use these resources to generate income. From an environmental perspective, this has the desirable spin-off that developing countries will now be less likely to engage in unsustainable development practices that are destructive of biodiversity.

National sovereignty over genetic resources has, therefore, been presented as a good thing, even a great victory, for the developing world, and indeed in many ways it is. After centuries of genetic resources being freely appropriated by scientists and profiteers from the richer countries of the world, in the 1990s it has become an accepted principle that a sovereign country has the right to control the outflow of germplasm from its territory and to be compensated when this occurs. The principle has led to a number of bilateral 'biodiversity prospecting' deals between Northern companies and Third World nations. The most famous of these deals is that between the pharmaceutical giant Merck and the Costa Rican non-profit NGO, INBio. In a two-year deal which has now been updated, Merck paid INBio an up-front sum of $1million to collect samples of biodiversity from the Costa Rican rainforest. They also transferred $135 000 worth of technology and are training Costa Ricans as parataxonomists. INBio will earn an estimated 1–5 per cent of any royalties on future products which result from the collected samples – which Merck will patent (Sittenfeld and Artuso, 1995). The money earned by INBio goes towards conservation of national parks and protected areas, and the agreement is fully supported by the Costa Rican government. INBio, undoubtedly the leader in such bilateral deals, also has arrangements with other companies, such as Bristol-Myers Squibb and the British Technology Group (Sittenfeld and Artuso, 1995; RAFI, 1996b, p. 49).

The reasons for the acceptance of the national sovereignty principle by Northern governments and transnational corporations are not hard to see. First, the deals based on national sovereignty are relatively cheap for the companies involved. The difference between the amount paid and the potential profits is vast. For example, under their updated deal, Merck pays INBio an upfront payment of $500 000 per year. They were also careful to ensure a good deal on royalties: the level set will depend on development costs and will probably be something like 1 or 2 per cent. Merck's annual R&D budget is $1.3 billion (Sittenfeld and Artudo, 1995); it costs Merck on average $125 million to develop a new drug. As RAFI point out, the money Merck is paying to Costa Rica, even if only one new drug arises from the prospecting mission, is 'barely loose change' (RAFI, 1996b, p. 49).

Second, national sovereignty over genetic resources merely replicates the relationship that predominates with regard to other raw materials the world over. It fits in nicely with the capitalist commodity relationship, creating a 'loop' of commodification in germplasm within which capitalist companies are only too happy to operate. Germplasm is bought from a (Third World) nation state, modified in the developed world by western scientists, and sold as a commodity under intellectual property rights.[2]

Third, national sovereignty does not threaten the intellectual property rights of Western companies. It may cost the companies concerned more than under the previous status quo (raw germplasm as common heritage; modified germplasm as protectable under intellectual property rights), but compared with the earlier solution advocated by the Third World (common-heritage-as-free-access to all germplasm, raw and modified) it is vastly preferable. Admittedly, Western companies – perhaps spurred by CBD clauses on technology and education transfer – have shown a willingness to share technology and educate local workers as part of the deals based on national sovereignty. This is commendable; nevertheless, it is clear that national sovereignty fits into the standard business way of doing things and is therefore happily accepted by the private sector.

SECTION II: NATIONAL SOVEREIGNTY AND COMMUNITY INTELLECTUAL PROPERTY RIGHTS

In Chapter 4 we argued on the basis of autonomy for the granting of legal rights to indigenous and farming communities in their traditional plant varieties and the knowledge they possess relating to genetic resources. Community intellectual property rights and the national sovereignty principle are, we believe, compatible. However, there is an obvious potential conflict between the two ideas. To understand the potential for conflict, we must go back to the principle of self-determination and its role in the relationship between indigenous communities and states in international law.

Self-Determination and Indigenous Peoples

The principle of self-determination is ambiguous in international law. As we noted above, some of the most important documents in international law grant self-determination expressly to 'peoples'. This immediately creates the problem of clarifying which groups are to be counted as peoples.

The meaning of self-determination, like that of other moral and legal norms, has evolved over time. It has been reinterpreted by successive generations to take account of the political realities of each era. The classical doctrine of self-determination, espoused most famously by USA President Woodrow Wilson at the time of the reorganization of Europe following the First World War, applies the principle to peoples in the sense of either ethnic groups or groups that have historically come under the rule of an accepted sovereign. Accordingly, while the polyglot Austro-Hungarian Empire was split into a number of small states along ethnic lines, the presence of a generally popularly accepted sovereign with a high degree of historical continuity meant that any division of the United Kingdom along ethnic lines was never considered.

In our own time the dominant way of thinking about self-determination, at least among the governments of the world, is very different. This conception sees the principle as applying very narrowly, to the populations of stable territories under sovereign governments. The explanation for this change in the dominant conception of self-determination lies in the period of decolonization following the Second World War. Colonial administrations were only very rarely based on ethnicity (Jackson, 1990, p. 77); indeed, they were more often based on the denial of ethnicity as the colonial powers pursued a policy of 'divide and rule' among the various ethnic groups in a given territory. For pragmatic reasons, post-war decolonization procedures had to accept colonial borders as they were; any reorganization of boundaries, particularly in Africa, would have been so complex and fraught with danger that it was unthinkable. Robert Jackson correctly states that '[t]he new state is the successor of an identical pre-existing European colony and is legitimate by right of succession regardless of the different ethnonationalities enclosed by it' (Jackson, 1990, p. 152). Under this conceptualization of self-determination, when international legal instruments talk of self-determination as a right of 'peoples', therefore, the term 'peoples' must be interpreted as the right of populations within stable territories, regardless of ethnicity.

This conceptualization has become a core doctrine of ex-colonial states themselves. The 1963 Charter of the Organization of African Unity affirms the equal sovereignty of members, the principle of non-intervention in their internal affairs, and mutual respect for the sovereignty and territorial integrity of each member and its inalienable right to independence (Jackson, 1990, p. 153). Jackson describes how the President of Mali emphatically confirmed the new doctrine at the first meeting of the OAU in 1964. The Mali President said that 'African unity demands ... complete respect for the legacy that we received from the colonial system' (Jackson, 1990, p. 153).

Fundamentally, then, the practicalities of international relations mean that the power of states in the international system, along with rights of sovereignty and self-determination, is generally taken to adhere to governments rather than ethnic groups. As James Crawford explains:

> [W]hen international law attributes rights to States as social and political collectivities, it does so *sub modo* – that is to say, it does so subject to the rule that the actor on behalf of the State, and the agency to which other States are to look for the observation of the obligations of the State and which is entitled to activate its rights, is the government of the State. (Crawford, 1988, p. 55)

However, in the last thirty years or so there has been an upsurge in international consciousness concerning indigenous peoples, their way of life and unique aspirations and needs. As a result of this rise in consciousness, various international agreements, draft declarations and conventions have been made which grant self-determination expressly to indigenous peoples. The most significant of these, the Draft UN Declaration on the

Rights of Indigenous Peoples, states plainly in Article 3 that '[i]ndigenous peoples have the right of self-determination', and that this includes the right to 'freely determine their political status and freely pursue their economic, social and cultural development' (quoted in Anaya, 1996, p. 209). However, any interpretation of self-determination that grants it to indigenous peoples is at odds with the dominant post-war interpretation that restricts it to populations under sovereign governments. For this reason, campaigners and writers seeking to gain self-determination for indigenous peoples have sought, first, to criticize the dominant interpretation, and, second, to forge a new doctrine of self-determination that takes account of changes in international perception of the claims of indigenous peoples.

A good example of this objective can be found in the work of Anaya. Anaya claims that indigenous peoples can be thought of as units of self-determination. The problem with the way self-determination is commonly interpreted by international policy-makers, he believes, is that it is identified too closely with sovereignty and full political autonomy and is therefore only applied to discrete and mutually exclusive 'sovereign' territorial communities. Anaya identifies three prominent versions of this interpretation of self-determination which he finds to be wanting. The first sees self-determination as only applying to colonized peoples. This version clearly derives from the linking of self-determination with post-war decolonization. To Anaya, this is too narrow; he believes self-determination is a principle derived from basic human rights and is therefore applicable to all peoples, not just a narrow set of colonized groups. Self-determination, furthermore, has an ongoing character and is not achieved once and for all when the colonial power is overthrown.

The second version of what might be called the 'Statist' view of self-determination, according to Anaya, argues that self-determination applies to the whole populations of independent states as well as to populations subject to colonial rule. However, the difficulty here is that only whole populations are thought to be entitled to self-determination, and not sub-state groups within the territory that may define themselves as peoples (Anaya, 1996, p. 77). Anaya wants to show that self-determination is applicable to indigenous peoples; he is therefore unhappy with this conception.

Anaya does not just criticize the dominant post-war interpretation of self-determination; he is also concerned with ensuring that the pre-war conception does not itself rise again. Hence he rejects the final version of the flawed common interpretation of self-determination, which does not apply the concept only to populations of stable territories, but applies it according to an 'alternative political geography' of ethnicity and historical sovereignty (Anaya, 1996, p. 78). According to Anaya, this overstates the importance of these factors and cannot explain the self-determination movements of the decolonization period in which ethnicity and historical sovereignty played little or no part.

The flaw all these conceptions share, believes Anaya, is that they divide the world into discrete and mutually exclusive 'sovereign' territorial communities. This worldview is that of classical western international jurisprudence, but it ignores the multiple, overlapping spheres of community, authority and interdependence that actually exist in modern human experience. The term 'peoples' must therefore be defined to accommodate the range of cross-cutting group identities real people actually have:

> Properly understood, the principle of self-determination, commensurate with the values it incorporates, benefits groups – that is, 'peoples' in the ordinary sense of the term – throughout the spectrum of humanity's complex web of interrelationships and loyalties, and not just peoples defined by existing or perceived sovereign boundaries. (Anaya, 1996, p. 79)

However, the problem with this characterization of self-determination is that, while it takes into account people's actual social, cultural and political experience, it evades more pressing political realities. Quite simply, decisions have to be made about which groups are to be accorded the right of self-determination, and in the interests of fairness, these decisions have to be made on the same basis in every case. As soon as these decisions are made, however, it seems that the world will be immediately divided into the discrete groups Anaya is so keen to avoid. Nowhere does Anaya demonstrate how the difficulties associated with such decisions can be avoided.

However, Anaya makes two other points that are relevant to our purposes and are more acceptable. The first is that wherever possible, we ought to interpret international law in as plain a fashion as possible; plain interpretation is the convention in international law and there ought to be no exception made when interpreting the word 'peoples' in connection with self-determination. When the UN Charter and other international legal instruments, therefore, speak of the right to self-determination as adhering in 'peoples', a plain interpretation suggests we should include indigenous peoples in our definition, and not just restrict the concept to whole populations.

The second important point made by Anaya is that self-determination does not necessarily include the right to independent statehood. The classical post-First World War doctrine of self-determination undoubtedly did see self-determination as including this right (and perhaps even as *entailing* statehood). However, as we argued above, we are not bound by the interpretations of the past, and must re-interpret norms to fit contemporary political realities and changes in consciousness. When documents such as the UN Draft Declaration on the Rights of Indigenous Peoples discuss the right of peoples to 'freely determine their political status', we can interpret this, as Anaya says, as being compatible with a variety of political arrangements, such as Trusteeship, 'associated statehood' – like that of the Cook Islands in relation to New Zealand – and regional autonomy – like that of the Austrians of South Tyrol (Brownlie, 1988, p. 6).

Indeed, indigenous groups, as Anaya writes, do not seem interested in secession or sovereignty as such: 'Group challenges to the political structures that engulf them appear not to be so much claims of absolute political autonomy as they are efforts to secure the integration of the group while rearranging the terms of the integration or rerouting its path' (Anaya, 1996, p. 79). Self-determination means that peoples should have the power to decide which political arrangements they want to be a part of. This might take many forms. Nevertheless, self-determination is often seen by international political decision-makers to involve full political autonomy; this is probably one of the reasons for the slow acceptance of the Draft Declaration on the Rights of Indigenous Peoples (Anaya, 1996, p. 111).

Self-Determination and Community Intellectual Property Rights

States that contain indigenous groups have often, perhaps understandably, seen these groups as a threat to their integrity. Such states include former colonial states that are made up of various ethnic groups, some of which might be described as 'indigenous'. These states are typically engaged in ongoing attempts to build a unified sense of national identity from their disparate groupings; and they have often looked unfavourably on groups that seek to maintain a separate identity from that of the mainstream. States have therefore often been reluctant to accord any special status to indigenous groups. Indigenous groups, however, have tended to resist the pressure on them to integrate and have repeatedly called for self-determination (Anaya, 1996, p. 75). The campaign for Traditional Resource Rights (TRRs)[3] can be seen as a manifestation of self-determination for indigenous groups.

Hence, national governments and the indigenous groups within their territories seem simultaneously to have rights of control over the same resources. There is quite clearly a fundamental incompatibility here. Unfortunately, looking to international law for clarity over this question fails to provide us with any assistance. *National* sovereignty over natural resources is, as we showed early on in this chapter, an unquestioned norm of international relations. Yet a number of international instruments also grant sovereignty to *peoples*. The International Covenant on Economic, Social and Cultural Rights, for example, links self-determination with the notion that 'all peoples may, for their own ends, freely dispose of their natural wealth and resources' (Crawford, 1988, p. 58). Even more emphatically, General Assembly Resolution 41/128 in 1986, the Declaration of the Right to Development, says that the 'human right to development ... implies the full realization of the right of peoples to self-determination, which includes', subject to the relevant provisions of both international covenants on human rights, the exercise of their 'inalienable right to full sovereignty over all their natural wealth and resources'. If we interpret the term 'peoples' to include indigenous peoples, as we have argued we ought to, there is an obvious conflict here.

This is not a tenable situation; national governments and indigenous peoples cannot both have sovereignty over the same resources. We believe that ultimately, sovereignty over natural resources must rest with states. This is the convention of international law and upon which the whole of the international system is based. States can be encouraged through various international fora to accord rights to their indigenous groups – and it is desirable that they grant large measures of self-determination to these groups – but sovereignty, and control over natural resources, lies with states. Overriding this principle carries with it the danger that the more powerful states of the world will use their might to go over the heads of the governments of developing countries to deal directly with indigenous groups. This would be a violation of one of the primary norms of international law and an act of hypocrisy. The governments of developed states would, after all, never allow their authority to be overridden in such a way. The rights over natural resources that go with the right to self-determination must, therefore, be circumscribed by national sovereignty.

However, this does not mean that community intellectual property rights are incompatible with national sovereignty. There is certainly no conceptual conflict: community intellectual property rights are conceived as a form of intellectual property rights rather than a form of sovereignty over tangible natural resources. And, crucially, in our conception, they apply to traditional varieties and botanical knowledge, not to raw natural resources whose value is not known. This is because, to recap, traditional varieties and botanical knowledge form a part of the culture of the group in a way that plants of unknown value do not. This entails a different set of legal rights from any set implied by a declaration of indigenous control over the natural resources of the inhabited area.[4] Community intellectual property rights imply a set of intangible property rights over a circumscribed set of resources and knowledge, while sovereignty implies physical control over the whole set of natural resources in a given territory.

This does not mean that national governments should allow bioprospecting to take place in the territories inhabited by their indigenous communities against these communities' will; the basic principle of self-determination implies that communities should be consulted before such activities take place in their territories. However, the principles behind community intellectual property rights do not mandate indigenous ownership of genetic resources of unknown value that happen to be found in their territories. In accordance with the principle of national sovereignty, benefits from the collection and use of such resources should accrue to the country as a whole and not just to local communities. This, naturally, includes local communities, and local communities should benefit just as much as the rest of the country in question. In fact, they may, in practice, benefit more, as the labour of local community members may well be needed by bioprospectors.

To implement this reconciliation between national sovereignty and community IPRs, an international regime would be required. The realities of international relations require that any such regime would have to be agreed by states alone; the indigenous groups of any state that chose to stay outside any *sui generis* system would not be covered. The question arises of whether it is likely that the governments of developing countries would enter into any arrangements that appear to devolve power over certain resources from them to minority groups within their territories. It may also be questioned whether any international arrangements concerning access to and transfer of genetic resources are likely to be made, given that, as Richard Falk points out, the 'statist character' of international arenas means that agendas and budgets are always controlled by states (Falk, 1988, p. 19).

However, the prospects of progress in international arenas in coming to an agreement over a *sui generis* system, have been improved by signs that the national governments of some developing countries are willing to protect the varieties and knowledge of their indigenous peoples. In the Philippines, for example, the Presidential Executive Order no. 247 requires the prior informed consent of both the government and indigenous cultural communities before access to any genetic resources is allowed. The Andean Pact (Bolivia, Colombia, Ecuador, Peru and Venezuela) has a 'Common System on Access to Genetic Resources' in order, amongst other things, to 'establish a basis for the recognition and appreciation of genetic resources, their derivatives and related intangible components, particularly where indigenous, afro-american and local communities are involved' (UNEP, 1996, p. 4). The 'Common System' requires secondary national legislation; in September 1996, Ecuador passed legislation which included a guarantee of 'ancestral rights over intangible knowledge and components, of biological diversity, and of genetic resources and control over them' (UNEP, 1996, p. 5).

Similarly, Brazil's Draft Bill of Law on Access to Brazilian Biodiversity requires the prior informed consent of local communities and also benefit-sharing that includes the participation of the country in economic, social and environmental benefits of any products and processes obtained through the use of Brazilian biodiversity (UNEP, 1996, p. 5). This law exemplifies an interesting symbiosis between the principles of national sovereignty and community intellectual property rights, and shows that these two principles can be quite compatible.

CONCLUSION

The full meaning of the national sovereignty principle in relation to genetic resources has not yet been decided at an international level. Because of the peculiar character of genetic resources compared to other types of resources, sovereignty and ownership can be conceived in a number of different ways. Genetic resources can be owned as tangible resources – that is, the actual physical manifestations of them can be owned – and as intangible resources, as information – through intellectual property rights. Furthermore, confusions over the meaning of the term 'peoples' in international legal instruments, and over the resource rights asserted on behalf of indigenous peoples in international instruments and fora, mean that exactly which group owns, or ought to own, genetic resources, is unclear.

National sovereignty over genetic resources has been presented in a number of different ways by various writers and commentators. Some believe it signifies that intangible – intellectual – property rights in germplasm should be granted to states (Margulies, 1993; Walden, 1995). Other writers believe indigenous groups ought to be granted intellectual property rights, while sovereignty should simply refer to tangible resources – that national sovereignty over genetic resources should mean the same as it does over other resources.

This lack of clarity means that any solution must reconcile not only the claims of indigenous communities and national governments, but also the different forms of ownership of genetic resources. We believe our suggestion – that indigenous communities, because of the need for special rights to protect their culture, should be granted intellectual property rights over their traditional varieties and botanical knowledge, while national governments should control the prospecting of all germplasm of unknown value – is a workable and satisfactory solution. There is even room for national governments to acquire a sort of intellectual property over the resources they control. They can use the national sovereignty principle to negotiate deals that allow access on condition that royalties are paid on any products developed out of a plant taken from the national territory.

The character of genetic resources is such that it took until the 1980s for the principle of national sovereignty to be applied to them. However, now it is an established norm. This are two reasons for this. First, as the developing countries of the world realized that the common heritage ideal, conceived as free access to all types of genetic resources, was dead, they also realized that an affirmation of national sovereignty was necessary if they were to avoid becoming suppliers of free raw materials to a Northern plant breeding industry that was using the intellectual property system to make a fortune for itself. The second reason was the increase in concern over the global environment, including the alarming rate at which the biodiversity of the subtropics was being destroyed. The Convention on Biological Diversity is the embodiment of the (quite correct) principle that if developing countries are to conserve their biological diversity, they need to be able to benefit economically. Without national sovereignty over genetic resources, countries have no way of receiving revenue from their biodiversity. National sovereignty is therefore a crucial component of any global framework for conservation.

However, the principle of national sovereignty is not the only principle that can benefit the developing countries. We have already seen how another principle, that of community IPRs, may benefit peoples. We must now turn to the common heritage principle, which is far from dead, and in a revised form may well benefit Third World citizens significantly in the future.

NOTES

1. In these days of environmental awareness, it is recognized that one state's use of its resources may have a deleterious effect on another state, and that these effects have to be considered by the state causing the harm. So national sovereignty does not mean a state can do whatever it likes with regard to its resources, regardless of the effects on its neighbours. Other norms intervene to prevent national sovereignty over natural resources becoming a *carte blanche*.

2. Developing countries can use the national sovereignty principle and the benefit-sharing provisions of the Biodiversity Convention to negotiate deals that involve not only the 'mining' of genetic materials for cash, but also involve training, technology transfer and other benefits. These deals are to be welcomed. Nevertheless, the new system still fits into standard business practice in that it treats germplasm as a national commodity to be exchanged for cash and other benefits.

3. 'Traditional Resource Rights' is the collective name given by campaigners to the whole swathe of rights connected to resources that campaigners believe ought to be accorded to indigenous groups. They include land rights and tenure rights, and also community intellectual property rights.

4. Although, if national governments want to accord such rights to indigenous groups, they are fully entitled to do so.

6 Common Heritage of Mankind

INTRODUCTION

If the principle of national sovereignty arises from the element of unilateral development in the Third World, with developing countries accepting the capitalist world economy and competing against one another within it, the principle of common heritage arises from the element of solidarity and multilateralism that exists in Third World politics. This chapter on common heritage is divided into three sections. In the first section, we outline our conception of common heritage. This conception is taken primarily from that expounded in the 1960s by the Maltese Ambassador to the UN, Arvid Pardo, with regard to the mineral wealth of the seabed. We elaborate and strengthen this conception by co-opting the 'difference principle' of John Rawls. In the second section, we show that, contrary to widespread belief, the common heritage ethic still retains an influence in the politics of genetic resource control. In the third section, we demonstrate that the common heritage ethic is compatible with both the community IPR principle and the national sovereignty principle.

SECTION I: THE COMMON HERITAGE ETHIC OUTLINED

It may seem odd to suggest that common heritage still has any meaning in the politics of genetic resource control. As an ideal, or as a way of organizing the use of genetic resources, common heritage appears to be dead. The phrase has even been dropped as a piece of diplomatic rhetoric from the International Undertaking on Plant Genetic Resources. However, common heritage is only dead as long as it is identified with common property and free access, and thought of as fundamentally opposed to intellectual property rights. This is indeed the common perception of common heritage in the area of plant genetic resources (see Chapter 2). Margulies, for example, writes: 'The compromise is not to negate all intellectual property rights in plant species, as would a common heritage principle ...' (Margulies, 1993, p. 345). In debates on the ownership of genetic resources, common heritage has typically been thought of as meaning 'common property'; i.e. referring to a situation whereby, in Drahos' terms, a regime of *res communis* applied to the commons of all genetic information. On this view, if common heritage were instituted, all germplasm, including elite lines, would be available to anyone without restriction. However, it is our argument in this chapter that this is not the only possible conceptualization of common heritage. Another, deeper, interpretation of common heritage exists and, indeed, still has an influence on the politics of genetic resource control. The best way to demonstrate this is to examine the roots of the idea of the common heritage principle.

Pardo and UNCLOS III

As an international legal concept, the principle of the common heritage of mankind (CHM) has its origin in debates on the ownership of the extra-territorial seabed. The phrase was first used by the Maltese Ambassador to the United Nations, Arvid Pardo, in 1968 to indicate his wish that the seabed outside territorial waters should not be exploited on a first-come-first-served basis, as this would favour the advanced nations over the developing nations by virtue of their superior technology. This would perpetuate global inequalities and was unfair, claimed Pardo (1968). The debates on ownership of the seabed, carried out through the Third United Nations Conference on the Law of the Sea (UNCLOS III), are instructive to us because they show that the conception of common heritage assumed by Pardo and his supporters was not one of free access or common property, but of common benefit. What was necessary was not that everyone should be given access to the resource, but that everyone on the planet should, as far as possible, benefit from its exploitation. To Pardo, this meant that the commercial benefits of seabed exploitation should go towards third world development (GA Resolution 2467A (XXIII) 1968). From the beginning, then, the concept of common heritage was not merely a property ownership claim, but a principle concerning the distribution of the benefits of a given resource. Alexandre Kiss called common heritage the 'materialisation' of the ideas of the common interest – 'that is, its application to material elements' (Kiss, 1985, p. 440) – while Bradley Larschan and Bonnie Brennan wrote of the UNCLOS negotiations that: 'the developing countries are attempting to implement a regime whereby they will receive the benefits of exploitation prior to their own ability to invest. This is the wellspring of the CHM' (Larschan and Brennan, 1983, p. 312).

Goedhuis provides us with the most systematic definition of the concept of the CHM. He draws a parallel between the UN Space Treaty of 1967 and the UNCLOS; the one establishing the whole of outer space, the other establishing the seabed, as the common heritage of mankind. Goedhuis identifies four elements in the concept of the CHM: first, the area in question – he conceives of common heritage in spatial terms, although this is not logically intrinsic to the concept, and we can replace the word 'area' with the word 'resource' without any loss of meaning – cannot be subject to appropriation. Second, all countries must share in the management of the area. Third, there must be an active sharing of the benefits from the exploitation of the area. Fourth, the area must be dedicated to peaceful purposes (Goedhuis, 1981, pp. 218–19).

Deepening the Common Heritage Ethic: the Difference Principle

The heart of the common heritage ethic, then, is the principle of common benefit. We can deepen our understanding of this conception of common heritage by using the ideas of John Rawls, though Rawls himself was sceptical about applying his theory of justice on an international scale. The relevant aspect of Rawls' theory of justice for our purposes is the 'difference principle.' Put simply, this is the principle of justice in distribution that affirms that economic inequalities are only justifiable insofar as they maximise the position of the least well-off. According to Rawls, this principle would be chosen by all rational agents in a hypothetical pre-political situation in which none of the agents knew their eventual position in society.

Rawls, in *A Theory of Justice*, did briefly apply certain aspects of his theory to the international arena – specifically those relating to justice in war and diplomacy (Rawls, 1971, pp. 377–82). However, he did not regard the difference principle as being applicable beyond the domestic boundaries of the state. As David Richards points out, Rawls employed a Humean conception of the proper sphere, or circumstance of justice, which sees justice as only applying when there is actual reciprocity – mutual advantage – between equal agents in a common arena (Richards, 1982, p. 276). Rawls sees the state as a cooperative venture between agents who are roughly equal – all members benefit from applying the rules of justice to each other and their institutions. In the international arena, however, there is neither a cooperative venture nor rough equality; nor is there mutual advantage or reciprocity.

Those who criticize Rawls for not extending the difference principle to the international arena, have taken two lines of attack. Charles Beitz, who has produced the most extensive neo-Rawlsian account of international justice, broadly accepts Rawls' theory of the proper circumstances of justice, but argues that the present-day economic interdependence of states means that Rawls' notion of a shared cooperative venture is present in the international sphere; we have the proper circumstances of justice and can therefore legitimately apply the difference principle internationally (Beitz, 1979, p. 151). Richards, on the other hand, argues that we do not need to rely on disputable facts like these to criticize Rawls' position – in fact, Rawls is simply mistaken in his views on the circumstances of justice. Morality does not involve actual, empirical reciprocal advantage but only involves reciprocity in the sense of role reversibility, or universalizability. All we need is the ability to see the other's point of view; and failure to do this is a failure of the moral imagination. Overcoming this failure has been at the heart of all the moral advances of the last few hundred years, such as the emancipation of slaves and of women, and the spread of democracy (Richards, 1982, pp. 277–78).

Among those neo-Rawlsians who have argued for the application of the difference principle to the international arena, it is disputed whether the parties to the original position would be individuals or states. When Rawls himself applied the original position to the diplomatic sphere, he envisaged the contractors as 'representatives of different nations who must choose together the fundamental principles to adjudicate conflicting claims among states' (Rawls, 1971, p. 378). In effect, then, the parties to the original position are states.

It is true that states are the main actors in international relations; however, this is not, we contend, actually the relevant issue in relation to international distributive justice. Individual persons are the parties to the domestic original position; and, as Beitz argues, the facts of global interdependence (or, *pace* Richards, mutual reciprocity) show the existence of a global scheme of social cooperation. This means that 'we should not view national boundaries as having fundamental moral significance ... Thus the parties to the original position cannot be assumed to know that they are members of a particular national society, choosing principles of justice primarily for that society' (Beitz, 1979, p. 151).

Rawls thought that a separate original position – and, therefore, a separate contract – for the international arena was necessary, and for such an original position, given that they were the 'actors', the contractors would be states. However, Beitz and Richards, albeit for different reasons, show that another original position, and another contract, is unnecessary. The knowledge of nationality must simply be denied to the individual contractors in the domestic original position. The domestic contract is thereby internationalized.

If the difference principle is, as has been argued, applicable internationally, how does it relate to common heritage and genetic resources? Within individual societies, Rawls' conception of justice as fairness leads him to a position of prima facie egalitarianism; only the difference principle can justify any move away from equality. Rawls arrives at this egalitarianism from his conviction that individual talents are arbitrary from the moral point of view. No one 'deserves' their talents, and no one 'deserves' to profit from them. Only if it benefits society's least well-off can such profit be countenanced (Rawls, 1971, p. 75).

In a similar fashion, neo-Rawlsian thinkers have argued that no country 'deserves' its natural resource endowments (Brewin and Donelan, 1978, p. 147), and that no individual deserves greater access to natural resources simply because she happens to have been born into a country with rich endowments (Beitz,

1979, p. 140; Richards, 1982, p. 290). In the original position, individuals would protect themselves against the possibility of ending up in a poor state by adopting the prima facie position that natural resources should be available to all in equal measure (Beitz, 1979, p. 141). Thus, from a Rawlsian position, all natural resources are, prima facie, common heritage interpreted as *res communis*. However, if a different system would maximise the position of the least well-off, this different system is preferable to one of *res communis*.

This is common heritage interpreted as common benefit, which, in our view, is the proper conception of common heritage. Here, Rawls' difference principle and the principle of common heritage come together: both are expressions of the common interest (Rawls, 1971, pp. 101ff; Kiss, 1985, p. 425). If we want to make more sense of this relationship for practical purposes, we can say that common heritage implies development of a resource for common benefit, and the difference principle tells us what common benefit – or the common interest – is: development in the interests of the least well-off. This ties in nicely with Pardo's original vision of common heritage for the resources of the sea bed: that it would be used for the benefit of humanity, which to him meant for the development of the Third World (Pardo, 1968, p. 135). We can see similar principles at work in the commitment to benefit-sharing found in the Biodiversity Convention.

SECTION II: COMMON HERITAGE NOW: BENEFIT-SHARING AND FARMERS' RIGHTS

The Convention on Biological Diversity, signed in 1992 at the Earth Summit, commits parties to ensure the sharing of the benefits of the new biotechnologies and the exploitation of the world's genetic resources that this entails. Article 1 calls for the 'fair and equitable sharing of the benefits arising out of the utilization of genetic resources', and in Article 19 (2) for 'access, on a fair and equitable basis ... to the results and benefits arising from biotechnologies based upon genetic resources provided by those Contracting Parties' (Grubb et al., 1993, pp. 76, 80). It would be hard to find a more representative example of the common heritage ethic. Crespi (1997, p. 232) refers to EU initiatives under the Concertation Action in the Biotechnology Programme (CUBE) to promote in cooperation with Third World countries the 'safe and sustainable exploitation of the renewable natural resources within their respective regions', whereby participating countries receive royalties generated by the commercial use of their resources. Crespi also points out that Material Transfer Agreements (MTAs) have been in place for several years, and that the FAO in 1993 produced an 'International Code of Conduct for Plant Germplasm Collection & Transfer'.

Benefit-sharing as a principle is specifically designed to avoid a situation whereby only those nations with the technological wherewithal to exploit genetic resources and benefit from them, do so. It promises a new relationship between North and South whereby the South exchanges its most valuable natural asset – its biodiversity – for northern technology and expertise, rather than simply selling it as raw material. As we have seen, Goedhuis' definition of common heritage includes as one of its main components a commitment to the active sharing in the benefits of the exploitation of a resource. Far from common heritage being dead as an ideal in the politics of genetic resource control, then, it is alive and well – albeit in a different guise.

One of the primary concepts the international community has developed to deliver benefit-sharing is 'Farmers' Rights'. The term 'Farmers Rights' first appeared in 1985 in the FAO's Commission on Plant Genetic Resources (CPGR). As Kloppenburg and Kleinman point out, the CPGR coined the term 'expressly to parallel the established concept of plant breeders' rights' (Kloppenburg and Kleinman, 1987, p. 36). It was essentially an attempt to stave off the ire of developing countries who, two years earlier, had pushed through the International Undertaking on Plant Genetic Resources, which declared all PGRs, including elite varieties, to be common heritage and therefore available without restriction. The situation demanded by the developing countries was never likely to come about; the industrialized countries were moving towards increased intellectual property protection for elite varieties and would never reverse that trend. 'Farmers' Rights' would be a way of giving the developing world some concessions without moving towards free access. As the CPGR saw it, on the one hand there would be plant breeders' rights providing legal intellectual property protection for elite breeders' lines, and on the other hand there would be 'Farmers' Rights' providing some sort of recognition of the contribution made to crop development and biodiversity by generations of third world farmers, and compensation for the use of 'landrace' germplasm in the elite lines of the industrialized world.

This was the original conception of Farmers Rights: 'recognition' plus a fund which would compensate third world farming communities for the use of their germplasm. This compensation would not link particular communities to any particular crop varieties, and could therefore in no way be considered an intellectual property regime (Frisvold and Condon, 1995, p. 45). It would, rather, take the form of an 'International Gene Fund' to which the plant breeding industry would contribute – voluntarily – whenever they used, in their breeding programs, a germplasm accession that originally came from the developing world. The Fund would be used to aid germplasm conservation programs (Shand, 1991, p. 135).

Nowadays the term 'Farmers' Rights' refers to a whole panoply of measures, some of which are being implemented and some of which remain mere suggestions, concerning the distribution, ownership and use of plant genetic resources around the globe. To get a flavour of the many varied suggestions, claims and demands that have been made under the banner of Farmers' Rights, we can look at a pro-Farmers' Rights article by Vandana Shiva published in 1996. In this article, Shiva makes several distinct Farmers' Rights claims. First, following the 'recognition' line, although in a rather ambiguous way, she claims that 'Farmers' rights reflect the recognition of sovereignty in ownership and creativity in traditional breeding by farmers' (Shiva, 1996, p. 1623). Second, she identifies a utilitarian element in Farmers' Rights, linking them to the conservation of biodiversity (Shiva, 1996, pp. 1623, 1631). Third, she maintains that Farmers' Rights include the following 'rights': the right to the informal innovations of farming communities; the right to save seed from one harvest to the next; the 'right to ecological security'; the 'right to food security'; and the 'rights' to produce 'diverse and nutritious foods for healthy consumption' and to 'challenge seed monopolies' (i.e., broad patent claims such as that of the company Agracetus on genetically engineered cotton and soybeans) (Shiva, 1996, p. 1631).

There is a danger that whatever anyone thinks would be a good thing for third world farming communities is included in their definition of Farmer's Rights. Shiva is not happy with Farmers' Rights being limited to the original formula of recognition plus a non-specific compensation fund. She finds that restriction 'inappropriate, insufficient and undignified', and complains that under this conceptualization, 'farmers do not have a place in negotiating biodiversity rights and determining patterns of biodiversity conservation' (Shiva, 1996, p. 1625).

A similar strategy to Shiva's is pursued by the Rural Advancement Foundation International (RAFI), a pressure group based in North America. In an article in the RAFI Communiqué, the author lists under the banner of Farmers' Rights such 'rights' as the right to land and secure tenure; the right to save seed and exchange germplasm; and the right of farming communities to refuse to make germplasm available to outsiders (RAFI, 1996a). These demands are often made and are perhaps warranted, but they illustrate the extent to which the rhetorical force of the phrase 'Farmers' Rights' has led organizations like RAFI to include any progressive demand connected to global agriculture under its name. 'Farmers' Rights' has, then, become a rallying cry for all progressive groups interested in global agricultural justice, and for third world governments interested in a better settlement for their agricultural sectors.

International policy-makers are quite confused about the meaning of Farmers' Rights. At the 1996 FAO Conference on Plant Genetic Resources for Food and Agriculture, delegates tied themselves in knots over the question. The ambiguity of the term 'rights' caused the most anguish. The USA doubted whether farmers could have any rights at all, given their vagueness as a group. The developing world complained about what they perceived to be the USA's insistence on individual rights, with Ethiopia saying that groups were more important than individuals and should therefore have more rights. The UK was more sympathetic than the USA to the idea that a group as vaguely defined as farmers could have rights, but argued that the top-down approach being taken at the conference was unlikely to have a beneficial outcome – 'grass-roots' initiatives were more desirable. India claimed that the EU's seat next to other countries at meetings like the present conference proved that a group could achieve legal recognition – as if the existence of the EU validated the whole normative concept of collective rights. Clearly, then, the use of the word 'rights' in relation to farmers causes a great deal of confusion, with representatives of different countries employing different conceptions of rights and consequently talking past each other.

Even the legal adviser at the 1996 Conference was not clear as to the meaning of Farmers' Rights. In his report on FRs, he stated that the concept of FRs signified a 'bundle of rights' with many strands. First, it was a call for the use of multilateral mechanisms and a fund to implement FRs. Second, it was a recognition of the 'farmer's privilege' – the right of a farmer to save seed from one harvest to plant in the next season. Third, it was a call for the legal protection of informal innovations carried out by farmers in non-developed countries. Fourth, it was a demand for a *sui generis* system for the purpose of protecting these informal innovations. Finally, it was a claim for compensation to farmers for the use of plant genetic resources they and their ancestors have developed over centuries of selection (ENB, 1997).

Only the first and last demands in this list are FRs' demands in the original FAO sense of the term. The core characteristics of the original conceptualization of FRs were that it was nonspecific and aimed primarily at benefit-sharing and conservation. The demands that are now made in the name of FRs concerning the institution of actual community intellectual property rights for informal farmer-based agricultural innovations and traditional varieties and knowledge are recent accretions to the notion of Farmers' Rights. They are different in kind in that they are specific rights given to specific communities.

The truth is that Farmers' Rights is not really a set of rights at all, but merely a set of suggestions, claims and demands connected to genetic resources and their distribution, use and ownership. The phrase 'Farmers' Rights' is therefore a misleading term that is causing a great deal of damage to the continued attempts of international policy-makers to try and reach a settlement over the issue. It ought to be dropped immediately. Let us explain why.

The phrase 'Farmers' Rights' was adopted for one reason and one reason only: because it offered a neat and diplomatic linguistic parallel with Plant Breeders' Rights. Plant Breeders' Rights are a well-defined system of legal arrangements that can accurately and legitimately be referred to as (legal) rights. When Farmers' Rights were originally mooted, agricultural sectors in the developing world, not to mention governments, were in a state of high agitation at what they perceived to be the unfair asymmetry between the application of Plant Breeders' Rights to elite strains of crop plants and the free availability of non-elite germplasm, which originates mainly in the developing world. To call a system that was aimed at assuaging the sense of injustice of the developing world a system of 'rights', was purely a political move. On the one hand, there would be Plant Breeders' Rights for the North; on the other hand, there would be 'Farmers' Rights' for the South. But the actual arrangements conferred no rights whatsoever on anyone; not developing states, and certainly not farmers. There were, as we indicated above, two components in these arrangements: first, a fund aimed at conservation of plant genetic resources, and second, a formal recognition of the role of the world's farmers in developing and conserving the world's genetic resources over the centuries. There are simply no *rights* conferred by such arrangements.

For there to be a right there must at least be a right-holder; perhaps we could designate the right-holder in the case of Farmers' Rights as 'the world's farmers'. But in addition to an assignable right-holder there must be, depending on the theory of rights being employed, either a beneficiary of the duty created by the right (the benefit theory of rights), or an assignable entity capable of exercising control of the performance of the duty (the choice theory of rights).[1] In this case there is neither. We suggest that we might designate 'the world's farmers' as the right-holder. But – to take the benefit theory of rights – it is not the world's farmers who would benefit from the original arrangements for 'Farmers' Rights', but the world in general, through the conservation of genetic resources. Now, it is true that sometimes the right-holder is not the same person as the beneficiary of a right (Jones, 1994, pp. 30–1; Steiner, 1994, pp. 59–73). This is the main argument made by proponents of the choice theory against the benefit theory. For example, when I promise my friend that I will help her to look after her ailing grandmother, my friend may then be said to have a right to my assistance, but it is not she who benefits from the performance of the correlative duty; rather, it is her grandmother. So perhaps it could be said that the farmers of the world hold the rights under Farmers' Rights, but the beneficiaries are everyone on the planet.

With this conception we have moved away from the benefit theory of rights to the choice theory. Under the choice theory, someone has a right when they are able to control the performance of another's duty. According to this theory, it is clear where the right lies in the case of my friend and her sick grandmother: with my friend, who may not be the beneficiary of the right, but who, after I have promised, has control over my assistance in that she can legitimately demand it at any time in the future. Like the benefit theory, the choice theory allows the right-holder and the beneficiary to be distinct persons. What is crucial, however, is that the right-holder has the power to control the performance of another's duty. But 'Farmers' Rights' in their original incarnation confer no such power on any assignable individual or group. The arrangements were aimed simply at conserving the world's genetic resources in the interests of present and future generations of human beings, as well as incorporating a moral recognition of the past role of third world farmers. They conferred no legal powers whatsoever; nor did they recommend the creation of such powers to national governments. There is no sense, then, under either of the two dominant theories of rights, that the original arrangements for 'Farmers' Rights' could be described as rights.

However, rights discourse is without doubt the most potent form of framing moral demands in today's world; when the word 'rights' is attached to a set of legal arrangements, its force is such that many other claims and demands immediately attach themselves to it. Since 'Farmers' Rights' were first mooted by the CPGR in 1985, a myriad of different demands and suggestions have been made under their rubric. These demands often have little in common with each other and are aimed at fundamentally different ends. But the power of rights discourse has led them to be lumped together with each other. Hence the current confusion surrounding Farmers' Rights in the international arena. In our view, the term 'Farmers' Rights' should be scrapped immediately and the suggestions made under its heading should be addressed individually according to their merits.

Lying behind some of these legal notions of Farmers' Rights is the view that farmers have moral, pre-legal rights that the law and international institutions and arrangements ought to 'recognize' or 'respect'. This view has been implicitly expressed in the international arena, with the 1991 revision of the International Undertaking on Plant Genetic Resources stating that Farmers' Rights 'arise out of the past, present and future contributions to the development and conservation of genetic resources that farmers have made throughout the centuries'. Such language may have diplomatic weight but it has no philosophical basis. The framing of political demands in terms of moral rights claims is rhetorically powerful but has merely increased the sense of confusion surrounding Farmers' Rights. The idea of 'Farmers' Rights' merely refers to a set of arrangements that may or may not be desirable for the conservation of genetic resources, and for North–South equity. The term can never indicate some kind of moral, pre-legal right that farmers, or farming communities, had and ought to be respected. When moral, pre-legal rights are said to 'arise' out of the performance of some act, then the entitlement theory of rights is being invoked. But as we saw in earlier chapters, the entitlement theory relies on both a distinguishable act and an assignable actor to which the resultant right can be said to adhere. The contribution of millions of farmers throughout history to germplasm development and conservation is not a distinguishable – to use Boyle's phrase, 'authored' – act, and 'the world's farmers' is far too vague a group to be the possessor of moral rights generated through labour.

It is true that some of the claims made under the rubric of Farmers' Rights could be accurately described as legal rights. The so-called 'farmers' privilege', for example, is in fact a legal right to save seed from one harvest to the next. Under the 1978 version of UPOV, the farmers' privilege still stands. However, under both patent law and the updated, 1991 version of UPOV, farmers are not permitted to save seed. While the farmers' privilege, where it exists, is undoubtedly a legal right, it is unlikely, therefore, that such a right will form part of a system calling itself 'Farmers' Rights'.

Although the original conception of Farmers' Rights did not include any call for the institution of a form of intellectual property rights for the informal innovation of farming communities, such rights have now become part of the collection of demands made under the term Farmers' Rights. Community intellectual property rights (CIPRs) are demands connected to Farmers' Rights that can accurately be termed rights. If instituted, they would grant the power to an identified community (or communities) to control the use, particularly the commercialisation, of a particular variety of plant, and – depending on the way CIPRs are conceived – to benefit financially from that commercialization. Under either the benefit or the choice theory of rights set out above, they can be counted as rights. On the benefit theory, the community benefits from the performance of the (legal) duty of others to obtain their permission before using the variety; and, on the choice theory, the community possesses the power to control others' use of the variety. CIPRs are therefore rights, in the legal sense. But as we argued in Chapter 4, even CIPRs are not moral rights. Certain writers have claimed that communities are entitled to rights in their traditional varieties because they have laboured to create them. We argued, however, that a community cannot be conceived as capable of labouring. A community cannot, therefore, generate moral (property) rights for itself. Nevertheless, we did argue, in Chapter 4, for the institution of CIPRs on the basis of autonomy.

'Farmers' Rights' – although a deeply unsatisfactory term – as reinterpreted in the light of the benefit-sharing provisions of the Biodiversity Convention proves that the common heritage ethic is alive and well in the area of genetic resources. The ideas that are actually grouped under the heading of Farmers' Rights are mostly valuable and would be useful in benefit-sharing and conservation. However, in our view, the term 'Farmers' Rights' ought to be abandoned. Its conceptual inaccuracy has generated so much confusion that no one seems to have any idea what it should mean. It has become an umbrella term for a variety of measures aimed at delivering benefit-sharing and conservation. Many of these measures are inspired by the common heritage ethic and are desirable in themselves; but their grouping under a title as indefinite as 'Farmers' Rights' has made their institution less likely. Delegates to the FAO conferences on the IUPGR have become embroiled in pointless bickering over the meaning of Farmers' Rights rather than engaged in serious debate over how to implement the many valuable ideas that come under its heading. This is the fault of the phrase itself; the sooner it is dropped, the better.

Our conclusion on Farmers' Rights, therefore, is that is it a confusing concept and that is serves no useful purpose in the debate over genetic resource control. All that is valuable in it can more usefully be expressed in either the principle of common heritage or that of community IPRs.

SECTION III: COMPATIBILITY BETWEEN COMMON HERITAGE, COMMUNITY IPRS AND NATIONAL SOVEREIGNTY

We want to demonstrate that the common heritage principle, as interpreted above, is perfectly compatible with both the principles of community IPRs and national sovereignty. First, let us deal with the relation between common heritage and community IPRs.

Common Heritage and Community IPRs

The common heritage principle, as defined above, means that the world's genetic resources should be used for the benefit of humanity, not simply appropriated by nations with the technological ability to exploit them. Applying the Rawlsian difference principle of distributive justice, this entails that control over genetic resources should be exercised to ensure that the position of the least advantaged groups in the world is maximized. Since, in general, the least advantaged groups in the world inhabit rural areas in Third World countries, the common heritage principle requires that the use of genetic resources must benefit them. Arrangements must be made, therefore, to safeguard and improve the situation of such groups, by requiring for example, adequate royalties to be paid by agricultural pharmaceutical companies to indigenous communities for the use of their genetic resources and traditional knowledge. More widely, to ensure that all disadvantaged communities share in the benefits derived from the exploitation of genetic resources, developed nations must be prepared to forbid transnational corporations (TNCs) from, for example, preventing farmers reusing seeds derived from previous years' crops.

Clearly, the principle of community IPRs is entirely congruous with this interpretation of the principle of common heritage. The claim that the autonomy of indigenous groups must be safeguarded by, inter alia, guaranteeing them control over their genetic resources naturally forms part of the strategy for satisfying the common heritage principle – in that one important way of maximizing the position of the least advantaged groups in the world is by protecting the cultural integrity of indigenous communities. Compensation by TNCs for past loss of genetic resources and traditional knowledge is one part of the deal; prevention of further loss in the future is another part of the deal.

Common Heritage and National Sovereignty

National sovereignty over genetic resources is often presented as antithetical to the common heritage ethic (Margulies, 1993, p. 332; Kloppenburg, 1988). But we argue that this is not necessarily the case. As long as common heritage is conceived in the way that the drafters of the 1983 International Undertaking on Plant Genetic Resources saw it, i.e. as free access to all types of germplasm, then the two principles are opposed. Clearly, the world's genetic resources cannot simultaneously belong to the whole of humankind and to nations; that is, something cannot be a global *res communis*, available to all without restriction, and also subject to the control of a nation state. But if we see common heritage in its fullest sense, of common benefit, then national sovereignty and common heritage are by no means incompatible. Indeed, national sovereignty is a mechanism by which we can achieve the aim of the common heritage ethic, i.e. the benefit of humankind in general from the commercial exploitation of genetic resources.

In certain areas, common heritage and national sovereignty are thought of not only as compatible but as mutually supportive. The United Nations Education, Scientific and Cultural Organization (UNESCO), for example, designates various monuments around the world as 'world heritage sites', noting their importance to humankind as a whole. But national governments nevertheless retain sovereignty over these areas – they do not come under the sovereignty of the UN simply because the UN has decided they are part of the global cultural heritage. There is no contradiction in this case – the two ideas, that these sites can be part of both the common heritage and also the national heritage, are eminently compatible. Genetic resources, of course, are different from monuments in that they are transportable and reproducible economic resources, but nevertheless, the principle still stands. In the area of genetic resources, Wood (interestingly employing the 'common benefit' conception of common heritage), has rightly argued that common heritage and national heritage are compatible:

> '[H]eritage' can be hierarchical: 'common heritage' and 'national heritage' are not mutually exclusive concepts. Crop germplasm can simultaneously be part of the common heritage of mankind, for the benefit of all, and part of a national heritage, to be exploited as a national resource. (Wood, 1988, p. 277)

At the end of the previous section, we argued that national sovereignty was desirable because it gave developing countries a lever in international law that they can use to ensure that they benefit from their primary natural assets. Clearly, then, national sovereignty can assist in attempts to achieve the aim of the common heritage ethic, which is the common benefit of all from the exploitation of the resource in question – in this case, genetic resources. Without national sovereignty, collectors from developed countries could collect germplasm from the developing world without having to pay, taking the collected resources back to the North where they would be developed into elite crop strains or pharmaceuticals using Northern technology and expertise, and would eventually come under the protection of intellectual property rights. With national sovereignty, developing countries can ensure they get a fair price for any collections made, and more importantly, can negotiate technology and skills transfers as part of access deals to ensure that they do not simply become raw material suppliers but acquire the expertise and infrastructure to develop their own resources in their own interests. Perhaps this is an idealized picture, but the point remains that the national sovereignty principle offers developing nations the opportunity to benefit from their genetic resources in ways that would be impossible without this principle being generally accepted. The common heritage principle, espoused and interpreted by Arvid Pardo in 1968, is aimed at preventing developed countries from using their technological superiority to capture all the benefits from a natural resource. National sovereignty over genetic resources will allow developing countries to develop their own resources on their own terms; it is therefore eminently compatible with the common heritage ethic as set out in this chapter.

CONCLUSION

In this chapter we have argued that the principle of common heritage can be shown to have continuing resonance in international debate over control of genetic resources. Transformed from its original meaning of *res communis* or everything in common, the principle of common heritage has come to represent the idea of common benefit, in which all communities, especially the least advantaged, are entitled to share in the benefits that can be derived from genetic resources. Interpreted along these lines, the principle of common heritage exemplifies the claims that are confusingly represented by the ubiquitous but misleading concept of Farmers' Rights. We argued that the notion of Farmers' Rights is unhelpful because of its intrinsic ambiguity and false attribution of rights, and that it ought to be replaced by the newly reinterpreted principle of common heritage. Finally, we demonstrated how the principle of common heritage is compatible with both the principle of community IPRs and the principle of national sovereignty. In our concluding chapter, we discuss the wider implications of these findings, and suggest some ways of implementing them.

NOTE

1.   For a discussion of alternative theories of rights (especially the benefit and the choice theories) see Jones (1994).

7 Conclusion

SUMMARY OF THE ARGUMENT

The arguments in this book are not intended to advance empirical claims about the causal relationship between ideas and practice, in the field of genetic resources. Our purpose is simply to clarify the basis of the normative ideas that lie beneath the surface of decision making on the issue of control over genetic resources. Whether normative ideas are independent of, or determined by, economic structures, the fact is that decision-makers, negotiators and campaigners must work with and discuss them. It is therefore essential that they are clear about the principles they are employing, and if these principles have a fundamentally moral basis, it is desirable that they have good arguments to underpin their moral convictions.

The politics of genetic resources is made up of four principles, each of which brings with it an ethic that its proponents believe ought to be taken up in fora where genetic resource control is discussed. These ethics come from different moral traditions, which is unsurprising, given that genetic resource control is a global issue and one in which most countries have some kind of stake. But since these moral traditions are at cross purposes with one another and often seem mutually contradictory, it is especially important that those involved in negotiating a settlement on genetic resource control are clear about the principles they are employing. We hope this book goes some way to clarifying the issues raised by these diverging principles.

Chapter 2 was an attempt to give an account of the present situation in genetic resource control. This was a difficult task given that so many aspects of the issue are not settled but are still the subject of negotiation and controversy. Nevertheless, we set out to explain, via the history of plant breeding in the twentieth century, why genetic resource control is a political issue. We argued that there are three main reasons. First and foremost, the event – albeit a gradual, historical 'event' – that turned genetic resource control into an issue of global significance was the rise in intellectual property protection for the products of plant breeding in the industrialized world, and in particular the flow of 'protected' germplasm from North to South. This led the developing world – the original source of most of the world's genetic resources – to move away from the principle of free access, after one last 'hurrah' for the principle in the 1983 International Undertaking on Plant Genetic Resources.

The second reason for the political salience of the genetic resource issue was the Third World's campaign for a 'new international economic order' in the late seventies and early eighties, which created a consciousness that the Third World was being exploited by the industrialized world. The third reason was the rise of biotechnology which alerted the world to the great potential value of genetic resources, both in monetary terms to TNCs, and in terms of their value to humanity as a whole.

In the next four chapters we examined the four main principles instantiated in the debates over genetic resource control. In Chapter 3, we discussed the most dominant principle – that of proprietarian intellectual property rights. In that chapter, our first task was to refute the fallacious argument that there is something fundamentally wrong with patenting living organisms. Since, in the chapter that followed, we argued for a kind of intellectual property right (community intellectual property rights) to living things, we needed to show that there was nothing wrong *per se* with the idea of IPRs. Our argument was that patenting does not, in itself, change humankind's relationship to nature in any way. Patenting merely signifies that certain living organisms can be used commercially by some people and not others. The view that actual, physical living organisms can be owned by people, is not controversial: owning organisms has been a universal human practice for thousands of years and is not seriously opposed by many people.

The remainder of Chapter 3 was taken up with a detailed analysis of the proprietarian interpretation of IPRs. Using the work of Drahos and Boyle, we described the way of thinking that currently dominates debates about intellectual property in the West, and, through GATT/WTO, globally. This is a way of thinking that rewards the type of information production in which there has been an act that can be interpreted as a 'first connection', or, to use Boyle's language, in which an author figure can be identified, whether that author figure be an individual or group. Proprietarianism ignores the negative effects of intellectual property on society, forgetting its own instrumental origins. With regard to the politics of genetic resources, the ethic of proprietarianism is particularly dangerous because the products of plant breeding and genetic engineering – elite crops and pharmaceuticals – are used worldwide. The negative effects of intellectual property protection for such products – which we described in Chapter 2 – are therefore global, but are not taken into account in settlements over intellectual property rights because of the dominance of proprietarianism within the WTO. This dominance is the result of the influence of the USA, which uses its extensive threat power to ensure that its strong proprietarian line on intellectual property is followed. However, this is to the detriment of many other countries, who neither produce protectable intellectual property on the same scale as the USA, nor possess indigenous industries strong enough to compete with US transnationals on their terms. An instrumental attitude to intellectual property, one capable of recognizing the detrimental effects of intellectual property and also of rewarding types of information production in which there is no author figure, and no 'first connection', would be preferable for most countries.

The fourth chapter dealt with the question of intellectual property rights for communities in their traditional varieties and knowledge. Indigenous communities are groups who possess and produce information but are discriminated against by the present system of intellectual property because it is difficult to identify an author figure or a first connection. Nevertheless, many people have argued that this information (in which we include traditional varieties, which contain genetic information 'produced' by communities over hundreds of years of artificial selection) ought to be protected. Many of their arguments are couched in entitlement language. However, we reject the idea that communities can be said to be *entitled* to intellectual property rights, because the requirements of an entitlement theory – that there be an identifiable act of creation, along with a creator – discriminate against communities in precisely the same way that proprietarianism does. A more satisfactory argument for community intellectual property rights is based on autonomy. We adopted Kymlicka's argument that cultural communities often ought to be granted special rights by virtue of the fact that they are endangered. These special rights include rights of control over various cultural forms, including traditional varieties and botanical knowledge. Granting these rights to communities gives them more control over their cultural forms and makes it less likely that such communities would wither away. This in turn means that the individuals in these communities would have a secure culture in which to develop themselves and their capacities and achieve and retain a sense of fully autonomous personhood.

In moving away from entitlement language, this argument aligns itself with the instrumental attitude towards intellectual property that we adopted in Chapter 3. Campaigners for community intellectual property rights have often used entitlement language because they believed that this would make their arguments carry more weight in intellectual property circles. However, information production by groups has always been ignored by the present author-centred or proprietarian entitlement mindset, because under this mindset, communities *cannot* be characterized as the potential holders of intellectual property. The entitlement mindset must be rejected if community intellectual property rights are to be accepted as feasible by those involved in the intellectual property discourse. The autonomy-based argument therefore goes hand-in-hand with an instrumental attitude to intellectual property.

In Chapter 5, we turned to the third principle of genetic resource control: national sovereignty. The principle of national sovereignty is very influential in the politics of genetic resources, because of its central place in the global system of states. Nevertheless, despite its global centrality, the principle of national sovereignty was not applied to genetic resources until relatively recently. It is true that genetic resources, along with other kinds of resources, were extracted from their colonies by the colonial powers, but mother countries were slow in exploiting them. Genetic resources are, crucially, information resources, but this fact was not recognized very quickly by either Northern countries or biodiversity-rich countries, even after the end of the colonial period. It was only recognized when northern countries began protecting elite crop varieties through intellectual property rights. Previously, biodiversity-rich countries had been happy to allow samples of their genetic resources to be removed from their territories free of charge. Now, however, national sovereignty is an accepted norm in genetic resource control.

We also argued in Chapter 5 that the principle of national sovereignty is compatible with the principle of community intellectual property rights as long as community IPRs are interpreted in the correct way. Correctly interpreted, the principle of community intellectual property rights implies a set of intangible property rights over a circumscribed set of resources and knowledge – traditional varieties and botanical folk knowledge. The principle of national sovereignty refers to the power to decide which arrangements ought to apply with regard to a particular resource, and it also implies control over a territory's wild genetic resources. The two principles can clearly coexist.

The sixth chapter examined the fourth and final principle of genetic resource control – that of the common heritage of humanity. We demonstrated, first, how the principle of common heritage has been transformed from the free-for-all notion of *res communis* to the morally idealistic notion of common benefit. Conventional wisdom would have us believe that 'utopian' ideals like common heritage are dead. Certain aspects of the politics of genetic resource control – such as the proprietarian victory in the Uruguay Round of GATT – seem to confirm this belief. Yet the common heritage ideal appears in the Biodiversity Convention's articles on the sharing of the benefits of biodiversity and biotechnology. And these articles are not merely rhetoric; states and companies are taking them seriously.

We pointed out how various agreements aimed at benefit-sharing have been made between donor countries and bioprospectors, and between states themselves. Of course, benefit-sharing can be interpreted in bilateral terms as a contract for the mutual benefit of the resource provider and the resource extractor, and such arrangements are not inspired by common heritage, which is the principle that all should benefit (or, in practice, that the developing world should benefit). Nevertheless, if we interpret the common heritage ideal in terms of Rawls' difference principle, it seems that most such arrangements are acceptable, as long as the position of the least well-off is maximized. More obviously, the common heritage principle can be seen in the various arrangements that come under the confusing title of 'Farmers' Rights'. In particular, the International Fund for Plant Genetic Resources, as a non-specific compensation fund aimed primarily at the conservation of plant genetic resources, but also at the conservation of farming communities, is a classic example of the common heritage ethic at work.

Finally, we demonstrated how the principle of common heritage is compatible with both the principle of community IPRs and the principle of national sovereignty. The combination of these three principles together constitutes a formidable challenge to the prevailing principle of proprietarian intellectual property rights, which currently dominates policy making on the issue of control over genetic resources.

FOUR IMPLICATIONS

There are four implications of this analysis. First, there is an implication for the legal system entailed by the story we tell in Chapter 2 concerning the spread of intellectual property rights to biotechnological inventions and the products of plant breeding and agricultural biotechnology. The legal system of a liberal democratic state is required to be impartial towards any particular interest of civil society. Yet this story indicates that intellectual property law has, over the last two decades, been consistently interpreted in favour of private interests, and that, as Drahos and Boyle show, the wider interests of civil society at large are being increasingly ignored. Perhaps this is inevitable, in that interpretations of the law necessarily reflect the most prominent ethic of the particular time in a society's history. Since the mid-1970s, in the industrialized countries – particularly the USA and the UK, which have the most proprietarian systems of all – the dominant ethic has been that of private enterprise and entrepreneurship. The proprietarian creed of intellectual property is the natural corollary of this wider ethic. It rewards conscious efforts at creating the type of information that can be easily commodified through intellectual property rights and sold for profit. This partiality in the legal system must be corrected.

The second implication concerns the Green movement, and arises from our refutation in Chapter 3 of the argument that there is something morally pernicious about patenting living things. The fact that the Genetics Forum finds patenting on living things 'repugnant' is not an argument against the practice. They have to show why. Significantly, once the Genetics Forum begins to justify this intuition, they immediately resort to consequential arguments, suggesting that it is not the patenting itself that is repugnant but the potential consequences. In our experience, this is characteristic of other deep green arguments; because of their general metaphysical proclivities, such arguments are often couched in intuitive terms, but once these intuitions begin to be explained and defended, reference to consequences are rarely far away.

The third implication concerns the relation between the principle of proprietarian IPRs and the principle of national sovereignty. In Section II of Chapter 3, we explained how the proprietarian ethic had spread across the globe because it coincided with the interests of powerful factions within the world's most powerful nation. This nation used its hegemonic position to force home the proprietarian ethic despite the fact that it was opposed by most of the world's less powerful nations. It did this by threatening these nations with trade sanctions if it did not get its way. However, the USA approached the GATT talks not only from a position of strength, but also in a mood of some anxiety about its position as hegemon. According to the USA, East Asian piracy of American intellectual property is largely responsible for the spectacular performance of East Asian economies in the past thirty years. This performance was threatening the interests of the USA and something had to be done about it. The USA used its strength to ensure that something was done, but some observers claim that its motivation was a perception that this strength was ebbing away, and that the GATT agreement of 1994 was not so much a signal of the continuing strength and authority of the USA, as the roar of a beast who senses that it is past its peak. Nevertheless, the combination of the principles of proprietarianism and hegemonic national sovereignty is a lethal force in policy making fora on genetic resource control.

The proprietarian ethic has undoubtedly spread worldwide because it is in the interest of the world's greatest power, and possibly some of the other rich countries of the world, that it does so. National sovereignty has become an accepted norm in the policy making fora, because it fits into the standard mode of natural resource extraction and therefore suits transnational corporations and the governments of industrialized countries. The proprietarian intellectual property ethic and the national sovereignty principle are the two most significant principles in the politics of genetic resources, in terms of the effect they have on policy. It is no coincidence that they each have powerful interests behind them and that they work effectively in harmony. Opponents of the proprietarian ethic must prise away from it the national sovereignty principle, and yoke that principle instead to the cause of community IPRs and common heritage.

The fourth implication is that the other two principles we identify in the politics of genetic resource control, despite the fact that they do not have any powerful *economic* or hegemonic interests behind them, do have considerable *political* clout. Community intellectual property rights are in the self-interest of the world's least-powerful group – indigenous communities – and yet they are being seriously discussed in international fora and may because the subject of a new *sui generis* intellectual property arrangement. Common heritage is the denial that self-interest should determine the control of genetic resources, or at least who benefits from them. Although we have argued that the common heritage principle arises out of the Third World multilateralist tradition, and therefore embodies a particular perception of the interests of the Third World, it is nevertheless a fundamentally moral principle in that it does not coincide with the short-term selfish interest of anyone in particular, but is aimed at the wider interest of all. Yet it still has some effect on political reality, as its inclusion in the Rio Convention on Biological Diversity indicates. What needs to be done is to consolidate the links between community IPRs and common heritage, and to wrestle the principle of national sovereignty away from its connection with the proprietarian principle. If the principle of national sovereignty were to support the case for community IPRs and common heritage (as it easily could), rather than to support the case for proprietarianism (as it does at present), a much fairer method of controlling genetic resources would be arrived at.

THREE POLICY RECOMMENDATIONS

The principal aim of this thesis has been to clarify the four principles that are present in the politics of genetic resource control. We believe this to be a fundamentally analytical undertaking. Conceptual vagueness has been responsible for much pointless wrangling in negotiations over genetic resource control, and for many incoherent arguments that have set campaigners' causes back. However, we have also interspersed the text with various more prescriptive policy-oriented opinions. Let us now recap on our three main policy recommendations.

Our first recommendation is a utopian one. It is that the common heritage ethic should guide all deliberations on genetic resource control. The aim should be that the world as a whole, and the developing world in particular, should benefit from any arrangements for genetic resources and the exploitation of the new biotechnologies. Of course, this will not happen overnight, since international agreements over economic matters are currently dominated by the self-interest of the world's most powerful nations. But by advocating the common heritage ethic, campaigners can at least ensure that the dominant proprietarianism is constantly challenged in international fora, and that the case for a more equitable way of controlling genetic resources is repeatedly heard in the public domain.

Our second recommendation is that the attitude of proprietarianism in Western conceptions of intellectual property is abandoned in favour of instrumentalism. Proprietarianism is detrimental to the interests of most of the world's countries and even to the wider public good of the industrialized countries. As many commentators have argued, it has led to a bias towards a certain type of research in biotechnology that will not necessarily in the long term give us the best inventions and products. Research aimed at producing commodifiable products that are protectable through intellectual property rights has been favoured at the expense of research that would lead to products that were directly aimed at fulfilling human needs or were less harsh on the environment. Proprietarianism has also led to indefensibly broad patents in biotechnological inventions that have the effect of deterring research, such as Agracetus/Monsanto's patents on all genetically engineered soybeans and cotton. A proprietarian system grants such patents because it sees its duty to reward, (in Drahos' terms) a 'first connection', or, (as with Boyle), 'authored' information production. Agracetus was the first company to produce genetically engineered cotton and soybeans, so it was given a patent on it. Under an intellectual property system that worked with an attitude of instrumentalism, and was capable of considering the wider effects of certain patents on the research community and society as a whole, such patents would never be allowed. To grant a patent on all genetically engineered plants of a certain species is to give too much power to a single company to direct all research on that species in its own interest rather than that of society.

Here, we are not advocating any changes in statute law. Intellectual property law as it stands is quite capable of considering the wider interests of society; an instrumental approach to intellectual property could therefore be adopted without any statutory changes. What is required is a shift in attitudes. Although such a shift may be difficult to achieve, we must be prepared to argue that the attitudes that currently dominate the intellectual property system are damaging and mistaken, and hope that these arguments are considered.

Note that abandoning proprietarianism does not entail abolishing patents on life. There is no reason to follow such a path; patenting of living things may have serious deleterious consequences in certain situations, but under an instrumentally minded intellectual property system, these could be taken into account and applications could be rejected on this basis. Furthermore, and perhaps most importantly, an instrumental system would allow each country to determine its own policy on intellectual property. The harsh consequences for the Third World of the post-GATT globalized proprietarian system could therefore be avoided.

Our third recommendation is that an international *sui generis* system be set up for community intellectual property rights in traditional varieties and botanical knowledge. Inaugurating such a system would undoubtedly present challenges, but there seems no reason to believe that they would be any more insurmountable than were the challenges of setting up a globalized proprietarian intellectual property system. Community intellectual property rights would preserve what is a vital part of many indigenous cultures and thereby help to promote individual members' autonomy. A *sui generis* system is necessary because the only alternative, a system of bilateral contracts, has too many drawbacks. A bilateral contract system would allow TNCs to play communities off against one another; it would hand too great an advantage to bioprospectors in what would be a 'buyers' market'; it would offer no recompense to communities in situations where outside interests accessed a community's variety or knowledge without permission or from elsewhere; it would mean that in disputes, communities would have to stand alone against the powerful companies they contracted with; and, perhaps most importantly, there would be no wider recognition that a particular variety or aspect of knowledge was the product of a particular community.

If these three recommendations were implemented, even in a diluted form, they would go far towards securing a just solution to the problem of how to control the world's genetic resources.

Bibliography

Anaya, S. James (1996) *Indigenous Peoples in International Law*, Oxford, Oxford University Press.

Bainbridge, David I. (1995) *Intellectual Property*, London, Pitman.

Becker, Lawrence C. (1977) *Property Rights: Philosophic Foundations*, London, Routledge & Kegan Paul.

Becker, Lawrence C. (1980) 'The Moral Basis of Property Rights', in Pennock, J. Roland and Chapman, John W. eds, *Nomos XXII: Property*, New York, New York University Press.

Beitz, Charles R. (1979) *Political Theory and International Relations*, Princeton, Princeton University Press.

Boyle, James (1996) *Shamans, Software and Spleens: Law and the Construction of the Information Society*, Cambridge, Massachussetts, Harvard University Press.

Brewin, Christopher, and Donelan, Michael D., eds (1978) 'Justice in International Relations', in *The Reason of States: A Study in International Political Theory*, London, Allen & Unwin.

Brown, William (1988) 'Plant Genetic Resources: a View from the Seed Industry', in Kloppenburg, Jack, ed., *Seeds and Sovereignty: the Use and Control of Plant Genetic Resources*, London, Duke University Press.

Brownlie, Ian (1988) 'The Rights of Peoples in Modern International Law', in Crawford, James, ed., *The Rights of Peoples*, Oxford, Oxford University Press.

Brush, Stephen B. (1993) 'Indigenous Knowledge of Biological Resources and Intellectual Property Rights: the Role of Anthropology', *American Anthropologist*, vol. 95, no. 3, pp. 653–71.

Brush, Stephen B. (1996) 'A Non-Market Approach to Protecting Biological Resources', in Greaves, Tom, ed., *Intellectual Property Rights for Cultural Peoples: a Sourcebook*, Oklahoma City, Society for Applied Anthropology.

Busch L., Lacy W.B., Burkhardt J. and Lacy, L.R. (1991) *Plants, Power and Profit: Social, Economic and Ethical Consequences of the New Biotechnologies*, Oxford, Blackwell.

Buttel, F.H. and Belsky, J. (1987) 'Biotechnology, Plant Breeding and Intellectual Property: Social and Ethical Dimensions', *Science, Technology and Human Values*, vol. 12, no. 1, pp. 31–49.

Cornish, W.R. (1993) 'The International Relations of Intellectual Property' *Cambridge Law Journal*, vol. 52, no. 1, pp. 46–63.

Cornish, W.R. (1996) *Cases and Materials on Intellectual Property*, London, Sweet & Maxwell.

Correa, Carlos M. (1992) 'Biological Resources and Intellectual Property Rights', *European Intellectual Property Review*, no. 5, pp. 154–7.

Correa, Carlos M. (1995) 'Sovereign and Property Rights Over Plant Genetic Resources', *Agriculture and Human Values*, vol. 12, no. 4, pp. 58–79.

Crawford, James, ed., (1988) *The Rights of Peoples*, Oxford, Clarendon Press.

Crespi, R. Stephen (1995) 'Biotechnology Patenting: the Wicked Animal Must Defend Itself', *European Intellectual Property Review*, no. 9, pp. 431–41.

Crespi, R. Stephen (1996) 'Patenting and Morality: A Patent Agent's Attempt at Moral Philosophy', *CIPA Journal*, June 1996, pp. 381–3.

Crespi, R. Stephen (1997) chapter 26 (no title) in Sterckx, Sigrid, ed., *Biotechnology, Patents and Morality*, Aldershot, Ashgate.

Crucible Group (1994) *People, Plants and Patents: the Impact of Intellectual Property on Biodiversity, Conservation, Trade and Rural Society*, Ottawa, International Development Research Centre.

da Costa e Silva, Eugênio (1995) 'The Protection of Intellectual Property for Local and Cultural Communities', *European Intellectual Property Review*, no. 11, pp. 546–9.

Dahlberg, Kenneth A. (1979) *Beyond the Green Revolution: the Ecology and Politics of Global Agricultural Development*, London, Plenum.

De Beof, Walter, Amanor, Kojo, Wellard, Kate and Bebbington, Anthony, eds, (1993) *Cultivating Knowledge: Genetic Diversity, Farmer Experimentation and Crop Research*, London, Intermediate Technology Publications.

Drahos, Peter (1996) *A Philosophy of Intellectual Property*, Aldershot, Dartmouth.

ENB (1997) 'Farmers' Rights', *Earth Negotiations Bulletin*, vol. 9, no. 47 (Internet journal: http://www.mbnet.mb.ca/linkages/vol. 09/0947011e. html).

Escobar, Arturo (1996) 'Constructing Nature: Elements for a Poststructural Political Ecology' in Peet, Richard and Watts, Michael, eds, *Liberation Ecologies: Environment, Development, Social Movements*, London, Routledge.

Falk, Richard (1988) 'Rights of Peoples (in Particular Indigenous Peoples)' in Crawford, James, ed., *The Rights of Peoples*, Oxford, Clarendon Press.

FAO (1983) *International Undertaking on Plant Genetic Resources*, Document C83/IIREP/4 and 5, 22 November, Rome, Food and Agriculture Organization.

FAO (1989) *Report of the Commission on Plant Genetic Resources*, Rome, Food and Agriculture Organization.

Fleising U. and Smart A. (1993) 'The Development of Property Rights in Biotechnology' *Culture, Medicine and Psychiatry*, vol. 17, no. 1, pp. 43–57.

Fowler, Cary (1995) 'Biotechnology, Patents, and the Third World' in Shiva, Vandana and Moser, Ingrun, eds, *Biopolitics*, London, Zed Books.

Fowler, Cary and Mooney, Pat (1990) *Shattering: Food, Politics and the Loss of Genetic Diversity*, Tucson, University of Arizona Press.

Frisvold, George B. and Condon, Peter (1995) 'The Convention on Biodiversity: Implications for Agriculture', *Technological Forecasting and Social Change*, vol. 50, pp. 41–54.

Gambles, Ian (1990) 'Global Distributive Justice: Rawls, Realism and the Priority of the Political Community' (unpublished paper delivered at the British International Studies Association Conference, Newcastle upon Tyne, December 1990).

Genetics Forum (1996) *The Case Against Patents in Genetic Engineering*, London, The Genetics Forum.

Goedhuis, D. (1981) 'Some Recent Trends in the Interpretation and Implementation of the Rules of International Space Law', *Columbia Journal of Transnational Law*, vol. 19, no. 2, pp. 213–33.

Greaves, Tom (1995) 'The Intellectual Property of Sovereign Tribes', *Science Communication*, vol. 17, no. 2, pp. 201–13.

Greaves, Tom, ed., (1996) *Intellectual Property Rights for Cultural Peoples: A Sourcebook*, Oklahoma City, Society for Applied Anthropology.

Grubb, Michael, Koch, Matthias, Thomson, Kay, Munson, Abby and Sullivan, Francis (1993) *The Earth Summit Agreements*, London, Royal Institute of International Affairs and Earthscan.

Hamilton, Neil (1993) 'Who Owns Dinner? Evolving Legal Mechanisms for Ownership of Plant Genetic Resources', *Tulsa Law Journal*, vol. 28, no. 4, pp. 587–657.

Hayek, Friedrich von (1944) *The Road to Serfdom*, London, Routledge & Kegan Paul.

Hegel, Georg W.F. (1967) *Hegel's Philosophy of Right*, trans. T.M. Knox, Oxford, Oxford University Press.

Hettinger, Edwin C. (1989) 'Justifying Intellectual Property', *Philosophy and Public Affairs*, vol. 18, pp. 31–52.

Hobbelink, Henk (1989) *Biotechnology and the Future of World Agriculture*, London, Zed Books.

Hohfeld, Wesley N. (1919) *Fundamental Legal Concepts*, New Haven, Yale University Press.

Honoré, Antony M. (1961) 'Ownership' in Guest, Anthony G. ed. *Oxford Essays in Jurisprudence*, Oxford, Clarendon Press.

Huft, Michael J. (1995) 'Indigenous Peoples and Drug Discovery Research: a Question of Intellectual Property Rights', *Northwestern University Law Review*, vol. 89, no. 4, pp. 1678–730.

Hughes, Justin (1988) 'The Philosophy of Intellectual Property', *Georgetown Law Journal*, vol. 77, pp. 287–366.

Jackson, Robert H. (1990) *Quasi-States: Sovereignty, International Relations and the Third World*, Cambridge, Cambridge University Press.

Jaenichen, Hans-Rainer and Schrell, Andreas (1993a) 'The "Harvard Oncomouse" in the Opposition Proceedings before the European Patent Office', *European Intellectual Property Review*, no. 9, pp. 345–7.

Jaenichen, Hans-Rainer and Schrell, Andreas (1993b) 'The European Patent Office's Recent Decisions on Patenting Plants', *European Intellectual Property Review*, no. 12, pp. 466–9.

Jayaraman, K.S. (1996) '"Indian Ginseng" Brings Royalties for Tribe', *Nature*, vol. 381, 16 May, p. 182.

Johnson, Lawrence (1991) *A Morally Deep World*, Cambridge, Cambridge University Press.

Jones, Peter (1994) *Rights*, London, Macmillan.

Khor Kok Peng, Martin (1990) 'The Uruguay Round and the Third World', *The Ecologist*, vol. 20, no. 6, pp. 208–13.

Kiss, Alexandre (1985) 'The Common Heritage of Mankind: Utopia or Reality?' *International Journal*, vol. 40, Summer 1985, pp. 423–41.

Kloppenburg, Jack R. Jr. (1988) *First the Seed: the Political Economy of Plant Biotechnology 1492–2000*, Cambridge, Cambridge, University Press.

Kloppenburg, Jack Jr. and Kleinman, Daniel Lee (1987), 'Seed Wars: Common Heritage, Private Property and Political Strategy', *Socialist Review*, no. 95, pp. 6–41.

Kymlicka, Will (1989) *Liberalism, Community and Culture*, Oxford, Clarendon.

Larschan, Bradley and Brennan, Bonnie C. (1983) 'The Common Heritage Principle in International Law', *Columbia Journal of Transnational Law*, vol. 21, no. 2, pp. 305–37.

Llewelyn, Margaret (1995) 'Article 53 Revisited: *Greenpeace* v. *Plant Genetic Systems NV'*, *European Intellectual Property Review*, no. 10, pp. 506–11.

Locke, John (1988) *Two Treatises of Government*, Cambridge University Press.

McKibben, Bill (1989) *The End of Nature*, New York, Doubleday.

Margulies, Rebecca L. (1993) 'Protecting Biodiversity: Recognizing Intellectual Property Rights in Plant Genetic Resources', *Michigan Journal of International Law*, vol. 14, no. 2, pp. 322–56.

Mill, John Stuart (1972) *Utilitarianism, Liberty, Representative Government*, London, Dent.

Mooney, Pat Roy (1979) *Seeds of the Earth: a Private or Public Resource?* Ottawa, Inter Pares.

Nachane, D.M. (1995) 'Intellectual Property Rights in the Uruguay Round: an Indian Perspective', *Economic and Political Weekly*, 4 February, pp. 257–68.

Nozick, Robert (1974) *Anarchy, State and Utopia*, Oxford, Blackwell.

Oasa, Edmund K., and Jennings, Bruce H. (1982) 'Science and Authority in International Agricultural Research', *Bulletin of Concerned Asian Scholars*, vol. 14, no. 4, pp. 30–44.

O'Neill, Onora (1986) *Faces of Hunger: an Essay on Poverty, Justice and Development*, London, Allen & Unwin.

Pardo, Arvid (1968) 'Who Will Control the Seabed?', *Foreign Affairs*, vol. 47, pp. 123–37.

Phillips, J. and Firth, A. (1994) *Introduction to Intellectual Property Law*, London, Butterworths.

Rabinow, Paul (1992) 'Artificiality and Enlightenment: From Sociobiology to Biosociality' in Crary, Jonathan, and Kwinter, Sanford, eds, *Incorporations*, New York, Zone Books.

RAFI (1996a) *Property Bioserfdom: Technology, Intellectual Property and the Erosion of Farmers' Rights in the Industrialized World*, Rural Advancement Foundation International Communiqué (Internet journal: http://www.rafi.ca.communique/fltxt/1996.html).

RAFI (1996b) *Conserving Indigenous Knowledge: Integrating Two Systems of Innovation*, A Rural Advancement Foundation International study commissioned by UNDP (United Nations Development Programme).

Rawls, John (1971) *A Theory of Justice*, Oxford, Oxford University Press.

Raz, Joseph (1986) *The Morality of Freedom*, Oxford, Oxford University Press.

Reid, Brian C. (1993) *A Practical Guide to Patent Law*, London, Sweet & Maxwell.

Richards, David A.J. (1982) 'International Distributive Justice' in Pennock, J. Roland and Chapman, John W., eds, *Nomos XXIV: Ethics, Economics and the Law*, Albany, SUNY Press.

Roberts, Tim (1994) 'Broad Claims for Biotechnological Inventions', *European Intellectual Property Review*, vol. 9, pp. 371–3.

Roberts, Tim (1996) 'Patenting Plants Around the World', *European Intellectual Property Review*, no. 10, pp. 531–6.

Rousseau, Jean-Jacques (1973) *The Social Contract and Discourses*, London, Dent.

Sahai, Suman (1994) 'Intellectual Property Rights for Life-Forms: What should Guide India's Position?', *Economic and Political Weekly*, 15 January, pp. 87–90.

Shand, Hope (1991) 'There Is a Conflict Between Intellectual Property Rights and the Rights of Farmers in Developing Countries', *Journal of Agricultural and Environmental Ethics*, no. 133, pp. 131–42.

Schapira, Ronald (1997) 'Biotechnology Patents in the United States' in Sterckx, Sigrid, ed., *Biotechnology, Patents and Morality* Aldershot, Ashgate.

Shiva, Vandana (1990) 'Biodiversity, Biotechnology and Profit: the Need for a People's Plan to Protect Biological Diversity', *The Ecologist*, vol. 20, no. 2, pp. 44–7.

Shiva, Vandana (1991) 'Biotechnological Development and Conservation of Diversity', *Economic and Political Weekly*, 30 November, pp. 2740–6.

Shiva, Vandana and Holla-Bhar, Radha (1993) 'Intellectual Piracy and the Neem Tree', *Ecologist*, vol. 23, no. 6, pp. 223–7.

Shiva, Vandana (1996) 'Agricultural Biodiversity, Intellectual Property Rights and Farmers' Rights', *Economic and Political Weekly*, 22 June, pp. 1621–31.

Sittenfeld, A. and Artuso, A. (1995) 'A Framework for Biodiversity Prospecting: the INBio Experience', *The Arid Lands Newsletter*, no. 37, Spring/Summer.

Smith, Adam (1976) *An Inquiry into the Nature and Causes of The Wealth of Nations*, Oxford, Clarendon Press.

Soleri, Daniela and Cleveland, David with Eriacho, D., Bowannie, F., Laahty, A. and Zuni Community Members (1996) 'Gifts from the Creator: Intellectual Property Rights and Folk Crop Varieties' in Greaves, Tom, ed. *Intellectual Property Rights for Cultural Peoples: a Sourcebook*, Oklahoma City, Society for Applied Anthropology.

Spector, Horacio M. (1989) 'An Outline of a Theory Justifying Intellectual and Industrial Property Rights', *European Intellectual Property Review*, no. 8, pp. 270–3.

Spencer, Herbert (1970) *Social Statics*, Farnborough, Gregg International.

Steiner, Hillel (1994) *An Essay on Rights*, Oxford, Blackwell.

Sterckx, Sigrid, ed., (1997) *Biotechnology, Patents and Morality*, Aldershot, Ashgate.

Subramanian, Arvind (1992) 'Genetic Resources, Biodiversity and Environmental Protection: an Analysis, and Proposals Towards a Solution', *Journal of World Trade*, vol. 26, no. 5, pp. 105–9.

Svatos, Michele (1996) 'Biotechnology and the Utilitarian Argument for Patents', *Social Philosophy and Policy*, vol. 13, no. 2, pp. 113–44.

Svatos, Michele (1997) 'Patents and Morality: a Philosophical Commentary on the Conference "Biotechnology, Patents and Morality"' in Sterckx, Sigrid, ed., *Biotechnology, Patents and Morality*, Aldershot, Ashgate.

ten Kate, Kerry (1995) *Biopiracy or Green Petroleum? Expectations and Best Practice in Bioprospecting*, London, Overseas Development Administration.

Tully, James (1980) *A Discourse on Property: John Locke and his Adversaries*, Cambridge, Cambridge University Press.

UNEP (1996) *UNEP/CBD/COP/3/20: Access to Genetic Resources*, New York, United Nations Environmental Programme.

Van Dyke, Vernon (1985) *Human Rights, Ethnicity and Discrimination*, London, Greenwood Press.

Verma, S.K. (1995) 'TRIPs and Plant Variety Protection in Developing Countries', *European Intellectual Property Review*, June, no. 6, pp. 281–9.

Walden, Ian (1995) 'Preserving Biodiversity: the Role of Property Rights', in Timothy Swanson, ed., *Intellectual Property Rights and Biodiversity Conservation*, Cambridge, Cambridge University Press.

Wells, Angus J. (1994) 'Patenting New Life Forms: an Ecological Perspective', *European Intellectual Property Review*, no. 3, pp. 111–18.

Winter, Gerd (1992) 'Patent Law Policy in Biotechnology', *Journal of Environmental Law*, vol. 4, no. 2, pp. 167–87.

Wood, David (1988) 'Crop Germplasm: Common Heritage or Farmers' Heritage?' in Kloppenburg, Jack, ed., *Seeds and Sovereignty: the Use and Control of Plant Genetic Resources*, London, Duke University Press.

WGTRR (1997) Traditional Resource Rights, Working Group on Traditional Resource Rights website (Address: http:// users. ox.ac.uk/~wgtrr/trr.htm).

Index

Note: references to 'genetic resources' 'plant genetic resources' and 'genetic resource control' have not been identified because they are too numerous to record.